Superviso ....
counsello
to the

GW01466130

Christina Hirst

RICS | the mark of
property
professionalism
worldwide

Published by the Royal Institution of Chartered Surveyors (RICS)
Surveyor Court
Westwood Business Park
Coventry CV4 8JE
UK
www.ricsbooks.com

First edition published 2004
Second edition published 2007
ISBN 978 1 84219 491 1

Typeset in Great Britain by Columns Design Ltd, Reading, Berks

Printed in Great Britain by Page Bros, Milecross Lane, Norwich

Contents

Foreword

In July 2006, RICS launched a revised APC, which incorporated many improvements to the competencies and guidance. This has been further enhanced with the upgrade of the APC templates in December 2008. This positive change is not only beneficial to candidates but is also designed to provide assessment panels with more concise information about a candidate's training and experience, allowing them to plan their interview questions more easily from the pre-submission documents.

RICS training adviser Christina Hirst looks at these changes in detail whilst retaining much of the helpful advice that was included in the previous editions of this publication.

I believe this book complements the official APC guidance produced by RICS and is a valuable resource for employers, supervisors and counsellors.

Suzanne Roberts

Membership Operations

Preface

When I was asked to update this book I felt this was a great opportunity for me to build on the excellent work of the original author, John Wilkinson, but also to take the opportunity to reflect the many changes that have taken place to the APC since its publication and to reflect current best practice. The aim of the book is to offer guidance for you in your role as supervisor or counsellor from the very beginning of your involvement with the APC to the success of your candidates at the final assessment. I would just point out that the objective of this book is not to replace the RICS formal guidance but to supplement this. The APC by its very nature must be dynamic and must adapt to reflect changes in the work and practice of surveyors; therefore, I would urge you to ensure that you follow the up to date guidance published by RICS at www.rics.org/apc. The contact details page at the back of the book provides further information.

I have been involved with the APC as an RICS training adviser for over ten years now and during that time have met a wide range of supervisors, counsellors and candidates working in a vast range of different organisations. There are a number of useful tips that all supervisors and counsellors seem to agree on:

- the importance of seeking advice on the development of a structured training agreement from an RICS training adviser;

- the need to put regular time aside to discuss APC progress with your candidate;

- the need to follow a timetable and action plan; and

- the need to adopt rigorous assessment of your candidate.

This book will help you address these and more.

The APC pathways and competencies have been developed and agreed by practising surveyors within the different professional groups. They are intended to reflect the work of surveyors within the particular specialisms. Generally you will find that a candidate's day to day work will allow him or her to meet the necessary competencies.

I cannot stress the need for structured training enough. The structured training agreement and any associated training plan gives you a process for managing the APC with your candidate. This should be a dynamic document used in monitoring progress and setting out the way by which the APC will be achieved. If things change, the plan should change accordingly. To set out on the APC without a training plan is like setting out on a journey when you know where you wish to end up but do not know how to get there. The outcome will surely be that you do not get there at all or if you do you will find the experience highly stressful and time consuming!

As a final comment, I would urge you to consider the personal development that you yourself gain by your involvement with your candidate: the development of supervisory, assessment and mentoring skills, to name but a few. This sounds like a good source of very cost effective lifelong learning or CPD to me!

Christina Hirst

Introduction to the APC

1

To understand what is required of you as a supervisor or counsellor, an appreciation of some of the background philosophies and key concepts of the APC is vital. This chapter is mainly designed for those of you who are new to the process – however, even those more experienced at APC training should find it useful. The APC is a constantly developing programme and all those involved must ensure that they have an up to date knowledge of its requirements and concepts.

Let's start at the very beginning.

WHAT IS THE APC?

The APC is the RICS' Assessment of Professional Competence. It is the practical training and experience, which when combined with academic qualifications leads to RICS membership. It is first and foremost a period of structured training and further study, and is followed by a 'final assessment' of each candidate's knowledge, experience and work to date.

The APC is undertaken by candidates who have graduated with an RICS accredited degree (but opportunities do exist to start the APC during part-time or distance learning study – see page 5) and who have commenced or are in relevant employment. Although supervisors and counsellors primarily only play a role in graduate routes 1 and 2, there are various 'routes' which will lead candidates to membership of RICS including:

- **Graduate route 1** – for those with an RICS accredited degree and less than five years relevant surveying experience;

- **Graduate route 2** – for those with an RICS accredited degree and between five and ten years relevant surveying experience (this may include pre-degree experience);

- **Graduate route 3** – for those with an RICS accredited degree and ten-years or more relevant surveying experience (this may include pre-degree experience);

- **Adaptation route** – for those with either an RICS approved professional body membership or non accredited but relevant degree and nine years relevant surveying experience. Candidates for this route must complete 450 study hours from an RICS accredited degree either at postgraduate level, from the final year of an undergraduate degree or from an approved training provider before taking the APC final assessment;

- **Academic** – for those who hold a surveying related higher degree and have undertaken academic activities on an RICS accredited degree over a three year period or who hold an RICS accredited degree and have undertaken academic activities relating to the profession over a three year period; and

- **Senior professional** – for professionals who are now in a **senior** industry position and hold a surveying related degree and have more than five years surveying experience.

The routes all culminate in attendance at a final assessment, albeit that the assessment for the senior professional route is different to other routes. Those who are successful at the final assessment gain professional status.

In essence, therefore, the APC is the process by which RICS seeks to be satisfied that candidates who wish to become members are competent to practise as chartered surveyors.

The assessment of competence is carried out over the full period of training or relevant experience – not simply at the final assessment interview. Supervisors and counsellors play a primary role in both the training and assessment of candidates.

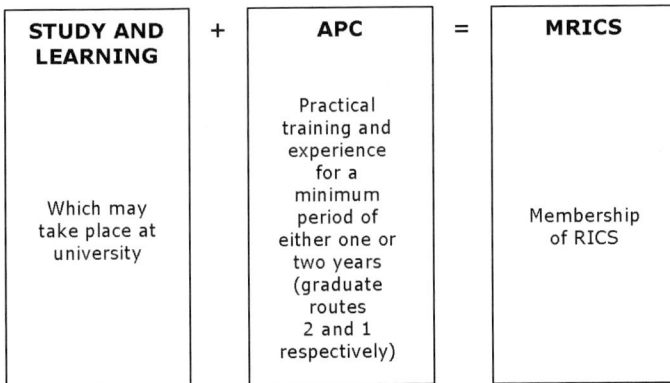

STUDY AND LEARNING	+	APC	=	MRICS
Which may take place at university		Practical training and experience for a minimum period of either one or two years (graduate routes 2 and 1 respectively)		Membership of RICS

Figure 1 Route to membership of RICS

The APC can be thought of in two parts:

- the structured training process; and
- the final assessment interview.

The structured training process

The structured training process should be rigorous and demanding. It takes place over a minimum period of 23 calendar months and includes a minimum of 400 days of relevant experience for graduate route 1 candidates or 11 calendar months and a minimum of 200 days for graduate route 2 candidates. The objective is to show that the knowledge and theory gained primarily in higher education is complemented with practical experience.

During the period of structured training, the candidate will need to gain a range of work experience that will allow them to achieve required and defined levels of competence for a specific APC pathway. The candidate's competence must be regularly assessed by a supervisor and a counsellor. Candidates are expected to keep a record of experience gained in a diary and to regularly summarise this in an experience record. They also need to undertake and record relevant professional development, and complete, along with the supervisor and counsellor, a candidate achievement record. These documents, and the role of the supervisor and counsellor with regard to them, are discussed further in chapter 5.

The final assessment interview

The final assessment interview is the second part of the APC process. A panel of a minimum of two assessors (although the panel will normally comprise three assessors) interview the candidate over a period of an hour and form a judgment, or 'assess', whether they consider he or she is competent to practise as a member of RICS. There are three main components to the interview: a critical analysis (submitted prior to interview); a presentation made during the interview on that analysis; and questions raised by the assessors exploring the candidate's knowledge and skills.

It is important to always bear in mind that the APC is primarily a period of training. Emphasis is placed on practical experience and it is this, as well as some theoretical knowledge, that is tested in the final assessment interview.

WHO CAN TAKE THE APC?

Candidates who have graduated with an RICS accredited degree and who have commenced employment or have relevant experience may take the APC. Part time or distance learning undergraduates can start structured training in their final year, and part time or distance learning postgraduates may do structured training concurrently with the last two years of their studies.

There are a number of routes as set out previously, but this book concentrates on graduate routes 1 and 2. Chapter 4 (pages 44 and 45) will be of assistance for supervisors and counsellors of graduate route 3 candidates but the book does not cover other routes nor routes to 'associate membership' (AssocRICS). However, the guidance in this book on supporting, supervising and counselling candidates will be of help to anyone carrying out those roles for candidates on those routes. In addition, the information provided on the philosophies behind RICS membership will be invaluable to anyone seeking membership – through any route.

For more information on any of the other routes to membership, readers should contact the RICS Membership Operations department or visit the RICS website at www.rics.org

WHO RUNS THE APC?

RICS is a membership organisation; decisions involving development or change are generally made by members sitting on various committees, on which RICS staff are also represented. The committee responsible for the APC in the UK is the UK Education and Standards Board. There are a number of such boards around the globe and their work is coordinated by the International Education Standards Board. These boards comprise groups of people representing the various national associations.

Policy developments that affect the structure of the APC are subject to approval by the relevant committee, and any changes to the technical competencies by the appropriate RICS professional group board.

Each national association will normally be involved in one of the Education Standards Boards, which represents the broad spectrum of membership in the country or group of countries that form the association.

The various professional groups (usually through their boards) will identify and define the pathways and competencies leading to membership of that professional group. This is an important concept as it demonstrates that the APC is developed by RICS members practising within that particular specialism. Some professional groups have more than one pathway to membership.

The APC is extremely dynamic. It is constantly changing to meet the requirements of the modern business world and a strategy of international development and growth.

OFFICIAL GUIDANCE

Official guidance on the APC can be obtained from the RICS website at www.rics.org/apc. There is guidance available for candidates and for those guiding and supporting the candidates. It provides a basic introduction to the various roles, together with guidance on the duties and responsibilities of the parties.

There is also additional guidance on the pathways, which provides contextual advice on the individual pathways from each professional group.

Further information on obtaining official guidance is provided in chapter 9 of this book. It should be stressed that the advice in this book is not 'official' guidance. Should discrepancies arise between the advice provided in this book and the official APC guidance, the official material should always take precedence. In any case, supervisors and counsellors should carefully read all the official guidance available and consult www.rics.org on a regular basis.

A last word of advice

The philosophies behind the APC are reflected in all aspects of the final assessment, and you will need to be aware of all the key points to guide your candidate through the process successfully.

2 Getting the right candidates

Before you can start supervising and counselling candidates through the APC, you must get hold of the candidates first. To attract the best candidates for your particular area of surveying, and geographical location, you will need to be able to offer them the best possible employment and APC package.

Recruitment of any new employee is an expensive business. Research shows costs of between £20,000–30,000 – expensive if you get it wrong. When trying to attract APC candidates, a good employer should start by considering whether all of the key ingredients are in place.

- Do you have a structured training agreement? (See chapter 5 for more information on this.) Most importantly, has this been approved by an RICS training adviser?

- What is your track record with previous graduates? Do you have pass rate percentages – and are they respectable enough to quote in job adverts?

- Do you have internal training modules and in-house programmes to support candidates on the APC and for their career development generally? Naturally, smaller firms may not be able to offer this but may be able to offer very focused support from supervisors and counsellors instead.

- Is your firm IIP (Investors in People) or ISO (International Organization for Standardization) accredited? This indicates a commitment to employees.

- What is your plan to help candidates meet the requirements of the APC? Will you be able to ensure that candidates will gain the necessary range of skills?

- Have you carried out any evaluation of your training scheme with your current graduates? If so, what have they told you about your performance?

- Do you have a training scheme for your APC supervisors and counsellors?

If you can answer 'yes' to all of the above questions, then you are ready to take on candidates. Being able to advertise the above will almost certainly help you to attract the best graduates, providing an instant return on your investment.

RECRUITMENT: A QUESTION OF 'FIT'

It is in everyone's interest to ensure that recruitment procedures select people that fit the organisation and job. The starting point in the recruitment of new staff is to consider what sort of person you are looking for, in the context of the culture of your organisation, your clients and customers, the team that the employee will be joining, and the skills required.

When advertising for employees, think carefully as to what information you can give about your organisation. The questions above show examples of the kind of things that attract the best APC candidates. You may also wish to offer visits to your office, so that candidates can experience the working environment first hand.

Working with universities offering RICS accredited degrees and offering opportunities for 'year out' placements for students and work experience over summer holidays is an excellent way to get to know potential recruits and to see if they 'fit' your organisation before entering into a long term commitment. For the student, this gives them a chance to gain some experience and for those on a year out placement to start their APC.

Application forms, interview techniques and assessments will all need careful consideration. Interviewing through the 'competency-based' approach (as in the APC interview) is a modern method, in which candidates are asked to give evidence of how they have employed their skills previously. This also fits neatly with the overall competency-based format of the APC, which I go on to discuss in later chapters.

Many large firms also hold 'assessment centre' selections to recruit graduate employees. In addition to a competency-based interview, other types of tests may be undertaken at these centres: aptitude tests, personality tests, group exercises and presentations.

A last word of advice

Recruitment and selection is a two-way process. To attract the best graduates and employees, you will need to impress them with a track record as an excellent trainer or potential trainer of APC candidates.

The role of the supervisor and counsellor

3

INTRODUCTION

When a candidate starts the APC process, he or she will
need to be assigned a supervisor and counsellor who are
usually from within his or her employer's firm. The
supervisor and counsellor will be chosen by the employer,
possibly in consultation with the human resources
department. In some cases, candidates may suggest their
own supervisor or counsellor, selecting someone they feel
can act as a mentor and provide good support within
their training area.

Supervisors and counsellors carry out a dual role: they
advise and support the candidate throughout the process,
and also assess the candidate's competence. They are not
expected to train a candidate entirely by themselves, but
rather, to make sure that the training the candidate
receives during the course of carrying out his or her
work in the firm matches and is applied to the
requirements of the APC. They must organise and direct
the training to ensure that this occurs, to avoid gaps in
the candidate's experience. They should involve the
candidate in all relevant work-based activities and
encourage colleagues to do the same.

Supervisors and counsellors must also assist the
candidate in complying with the requirements of the APC
in terms of submitting material and filling out
documentation, and must prepare them for the final

assessment interview. Vitally, they are the people who assess the candidate as having met the necessary levels of each required competency and ultimately confirm that the candidate is ready for the final assessment.

The supervisor should be the person who has day-to-day responsibility for the candidate and a good knowledge of his or her work. The counsellor has a more strategic role to play; he or she is responsible for planning the training programme and monitoring the progress of the candidate on an overall, 'big picture', basis, ensuring that the programme meets with the requirements of the candidate's chosen APC pathway.

The supervisor does not need to be RICS qualified, but the counsellor must be a chartered surveyor. Preferably, both should be members of RICS and should come from the same pathway that the candidate is following to membership. In practice, RICS will be comfortable with a non-chartered person carrying out the role of a supervisor, so long as they are suitably qualified to give guidance to the candidate concerning training and day-to-day work. Note also that it is acceptable for the counsellor to come from a different firm – a useful provision if there is no-one in the candidate's own firm able to fulfil that role. In this situation a process will need to be put in place to develop a mechanism to allow the counsellor to understand the work with which the candidate is involved and to assess their competence in undertaking that work.

There are significant differences between the roles of the supervisor and counsellor, and RICS would ideally wish the roles to be carried out by two different people. This allows the candidate access to two sets of knowledge, and should provide for a more impartial experience. However, in smaller firms, and under certain sets of circumstances, one person may be obliged to combine the two roles. If this happens, that person must be a member of RICS.

The roles are quite a heavy responsibility. In a moment we will consider the duties of both parties, in chronological order, from enrolment through to the final assessment – and beyond. Before that, it may be useful to step back and consider the overarching responsibilities for the entire training period.

In brief, as a supervisor or counsellor, you must:

- ensure, day to day, and week to week, that your candidate receives training in line with the competency requirements of his or her APC pathway;

- deliver coaching and training yourself, or ensure that the candidate receives training with someone else, for example, within another department in your firm;

- assess the candidate at three-monthly and six-monthly intervals throughout the APC training period;

- ensure that all records and reviews are completed accurately, reflect the requirements of the APC, and are submitted on time;

- assist the candidate with the preparation and submission of documents for the final assessment, and in preparation for the final assessment interview;

- liaise regularly with the supervisor/counsellor (if you are not the same person); and

- provide support and encouragement, and generally be a good friend to the candidate, including on matters of general health and welfare. Consider yourself perhaps more as a mentor for the candidate – someone they can bounce ideas off as well as be assessed by.

The most important thing to realise is that your involvement with the training will be continuous, across

the full training period – as will the process of
assessment. You will not simply be 'dipping in and out'
of the process.

WORKING TOGETHER

While not carrying out exactly the same roles,
supervisors and counsellors should look to form an
active partnership in terms of their respective
responsibilities. This allows them to support one another,
and provides a consistent management approach for the
candidate. Under no circumstances should a situation
arise in which the supervisor and counsellor are working
against one another.

The active partnership should take the form of regular
discussions and communications, outside of the formal
six-monthly intervals, to head off any potential problems
and ensure that the formal reviews run smoothly.

CHECKLIST OF ACTIONS

Having taken account of the overarching responsibilities,
we can now have a look at the nuts and bolts of the
process, from the point of view of the supervisor and
counsellor. The brief notes below give some ideas of the
duties to be performed at various stages of the process to
the final assessment. You should treat this as a handy
checklist – nothing more. The detail of the duties is
discussed further in chapter 4.

Bear in mind that the following is 'an ideal world'
scenario. Business or other demands may prevent your
candidate from completing the APC within the two years
(or one year if graduate route 2). If that is the case then
the training period will simply be extended and the
regular assessments maintained.

Month 0: enrolment

As an employer your involvement will actually start even before enrolment. Before the candidate enrols you will need to ensure that the organisation has a structured training agreement in place that has been approved by an RICS training adviser (RTA). Page 113 gives further advice on the role of RTAs and contact details can be found at www.rics.org/apc. For the candidate, enrolment is really the start of the APC process. Prior to the candidate formally enrolling, the supervisor and counsellor must meet with the candidate and ensure that the structured training agreement is in place, choose the most appropriate pathway and competencies for the candidate and make sure all relevant documentation is sent in to RICS promptly.

A delay of a few weeks of the candidate enrolling could put the final assessment back six months, as final assessments are normally held twice a year. Advice on closing dates are given at www.rics.org/apc

Supervisor and counsellor actions

- Meet with the candidate and explain the respective roles.

- Ensure that a structured training agreement is in place and approved by an RTA.

- Develop a training programme to forward plan the candidate's training to ensure the competencies and levels of the chosen pathway will be met with a timetable for each stage of this process.

- Check that the candidate returns all of the appropriate documentation to RICS in a prompt and timely fashion.

- Put dates in your diary for the three- and six-monthly reviews and make sure these are agreed between all parties required to attend.

- Agree a target final assessment date and build this in to the timetable.

Three-monthly supervisor reviews

Every three months, the supervisor must carry out an assessment of the candidate's competence. The candidate should present their diary, log book and experience record in advance of this assessment and the supervisor should assess the candidate's performance against each competency and advise the candidate of the levels achieved. The supervisor's progress report (an example form can be found at www.rics.org/apc) can be used to provide feedback to the candidate. A review of the next stage of training should also be made.

Supervisor actions

- Check that the candidate's training is progressing as planned: review the structured training programme and progress against the various competencies of the chosen pathway.

- Review the candidate's diary, log book and experience record and advise the candidate of any competency levels that have been reached. Ensure the candidate records this in the log book summary in the candidate achievement record (template 3).

- Review the professional development record (template 4), to ensure the candidate is making good progress in this area.

- Check the candidate is filling in the diary (example form at www.rics.org/apc) and the log book within the candidate achievement record (within template 3) as required.

- Complete the supervisor's progress report (example form at www.rics.org/apc).

Counsellor actions

Although you have no formal role to play at this stage, for reasons of good management it is desirable for you to speak to the supervisor, to ensure that everything is on track. That way, if problems have arisen, or are likely to arise, efforts can be made to head them off early, rather than waiting for the more formal review process – by which point minor issues may have become major difficulties.

Supervisor and counsellor six-monthly reviews

At the end of every six months there is a further review, in which the counsellor is involved. The aim is similar to that of the three-monthly review: to ensure that the candidate is progressing well against the competencies and that documentation is up to date.

Supervisor actions

- Check that the candidate is progressing with the training as planned: review the training programme and progress against the various competencies of the chosen pathway.

- Review the candidate's diary, log book and experience record and advise the candidate of any competency levels that have been reached. Ensure the candidate records this in the candidate achievement record.

- Review the professional development record, to ensure the candidate is making good progress.

- Check the candidate is filling in the diary, log book and experience record regularly as required.

- Complete the supervisor's progress report.

Counsellor actions

- Review the candidate's training and achievements to date.

- Review the candidate's diary, log book and experience record

- Assess the candidate's performance against the competencies, advising the candidate of those that you feel have been achieved. Ensure the candidate updates the candidate achievement record to reflect your assessments.

- Complete the counsellor's progress report.

At the end of 12 months for a graduate route 1 candidate or at the first three-month assessment for a graduate route 2 candidate both the supervisor and the counsellor should undertake a full review of the training programme and discuss possible projects that the candidate might use for their critical analysis. Progress on this should then be discussed at each of the next assessments. An action plan should also be put in place to help the candidate gain any additional experience needed, prepare for final assessment and plan for their additional reading and revision.

At the end of minimum 22 months (minimum 11 months for graduate route 2)

This is the earliest point at which a candidate can apply for the final assessment. See page 41 for more advice on things to consider in deciding whether the application should be made at this point.

One month after the application has been submitted for final assessment

At this point the supervisor and counsellor will need to ensure that the candidate submits the necessary documentation to RICS, ready for the final assessment. After submission the candidate will receive a letter from RICS confirming the assessment centre, date and time (normally one moth prior to the assessment period).

Supervisor actions

- Confirm any outstanding competency achievements and ensure the candidate records these in the candidate achievement record.

- Review the professional development record to ensure that a satisfactory number of hours have been achieved over a range of activities.

- Ensure the log book is up to date and shows compliance with the minimum number of required recorded days.

- Check the critical analysis.

- Check the experience record.

- Complete the declaration on template 1 of the candidate's submissions.

- Ensure the candidate submits the necessary documentation to RICS on time. Thoroughly check that this documentation is complete and accurate; an incomplete submission will be returned by RICS and final assessment could be delayed by six months.

- Offer advice to the candidate on the final assessment interview.

- Discuss a revision plan with the candidate.

Counsellor actions

- Confirm any outstanding competency achievements and ensure the candidate records these in the candidate achievement record.

- Double-check the professional development record to ensure that a satisfactory number of hours have been achieved.

- Check the critical analysis and experience record.
- Double-check that the log book is up to date.
- Offer advice to the candidate on the final assessment interview.
- Discuss a revision plan with the candidate.

A last word of advice

Supervisors and counsellors must work together; your candidate needs a coherent management approach in order to progress with confidence.

In chapter 4 all of these stages are discussed in detail. The documentation referred to is discussed further in chapter 5.

4

In detail: supervisor and counsellor duties

The previous chapter discussed the roles of the supervisor and counsellor in brief, giving a short checklist for the key activities. This chapter aims to 'flesh out' those checklists and notes, to give you a fuller understanding of what is required from you in your roles.

At each of the stages noted below, the need to follow the competencies set out in the APC guidance is mentioned. It is essential that supervisors and counsellors have a firm grasp of what exactly the competencies are – and of how they should be measured and assessed. The competencies are discussed – and explained – in detail in chapters 6 and 7.

Although the roles can be divided up into stages, the APC is a process of continuous training and assessment – the supervisor (and to a lesser degree, the counsellor) does not simply dip in and out of the training at certain specified dates. Your involvement with the candidate's progress is continuous, and your assessment and training plan must be continuous and progressive also. It may therefore be helpful to start with a quick look at the nature of the assessment process overall.

CONTINUOUS, PROGRESSIVE, COMPETENCY-BASED ASSESSMENT

Continuous assessment

The APC process is one of continual training and learning. It is therefore one of continuous assessment too. Although the final assessment will be carried out by trained APC assessors, supervisors and counsellors are involved in the ongoing assessment of candidates over the full course of the two-year training period. You are called upon to make formal judgments on the competence of candidates at certain specific stages, to certify that they meet the standards set by RICS.

If you are the candidate's supervisor, it is likely that you will already be doing much of what is required to assess him or her, as this requires similar processes to those used for appraising staff. Assessment involves being aware of how the candidate is performing in day-to-day activities and looking at work he or she has produced. From this, you can begin to form a judgment of how well he or she is doing.

If your candidate is in another department, or on secondment, you will have to work a little harder. By observing your candidate at work in that department, you will be able to assess some of the requirements of the mandatory competencies, such as working in a team, problem-solving and working to deadlines. By looking at work the candidate has produced, you will be able to learn more about his or her technical and professional knowledge, as well as his or her understanding. If you are an external counsellor (in other words, not working in the same organisation) you will need to arrange to see examples of the candidate's work and perhaps to discuss this with some of their colleagues.

Obviously, you should keep notes and records of how your candidate is doing, to help you with the 'signing off' of the competencies. Keep such notes regularly, not just when a formal review is approaching, to give you a thorough overview of his or her performance.

Progressive assessment

One of the basic philosophies of the APC is the idea that candidates will learn skills and acquire experience progressively. This is embodied in the levels set out for the various competencies, in which candidates are expected to progress from a 'knowledge and understanding' of a particular area, through to practical experience of that area under normal circumstances, and finally, to gain the ability to apply their knowledge in more complex circumstances, and to evaluate a situation in a wider context.

The competencies are drawn up with three levels to reflect this progressive approach. Chapter 6 considers this in some detail – meanwhile, supervisors and counsellors should bear in mind, as noted throughout this chapter, that the candidate's records must show a progressive approach to learning. Supervisors and counsellors do not wait until a candidate has reached level 3 before confirming achievement – but rather indicate how they have progressed through levels 1 and 2, in earlier records and reviews, to reach this higher level. The final assessment panel will take a dim view of all competencies being confirmed on the same day, as this implies that the progressive approach has not been adopted – or at least, not properly monitored.

In the APC, the process is as important as the end point.

Competency-based assessment

The APC consists of competency-based assessment. This is assessment of actual performance, or competence, at work. It seeks to ascertain that the candidate not only has the necessary knowledge and understanding to carry out his or her work, but can also put this knowledge into practice. Traditional qualifications tend to test what people know, whereas the competency-based approach, while still assessing what they know, also assesses what they can do.

However, it should be recognised that at certain points in the final assessment interview, the assessors may 'step outside' of the candidate's experience, and ask him or her wider questions on a particular issue. This type of questioning is discussed in more detail on page 45.

Your assessment and reviews should prepare the candidate for this type of approach. First, ask your candidate to provide you with examples of practical occasions on which they have put their knowledge into practice. Ask them to explain what approach they adopted in a particular instance, and to analyse the effectiveness of that. Then, try to extend this experience, asking them about factors surrounding the actual experience – potential problems that could arise another time, for example. Again, see page 45 for more information on this type of questioning.

We will now look at the roles of the supervisor and counsellor in more detail, starting at the very beginning of the APC process.

ENROLMENT

Before your candidate can enrol for the APC you will need to ensure that the organisation has a structured training agreement in place that has been approved by an RICS training adviser (see page 55). This is a crucial document as it sets out in detail how the candidate will be supported to achieve the requirements of the APC training and experience.

Enrolment is the start of your APC candidate's journey to membership. It will also form an important part of your candidate's first impressions of you (as supervisor or counsellor), and your organisation.

Ideally, supervisors and counsellors hold a short meeting to brief their candidates on the APC process, provide an overview of what it entails and introduce the organisation's structured training agreement and principles. They should stress the importance of completing the enrolment forms as soon as possible. The sooner your candidate is enrolled by RICS, the sooner the candidate will be able to start recording their experience and training – and the sooner he or she will be able to consider the final assessment. More information on the documents to be used for this is given in chapter 5.

The supervisor and counsellor must also discuss and agree the competencies to be achieved, considering the candidate's chosen pathway to membership, and the resources of the firm. (The competencies, and how to choose them, are discussed in detail in chapters 6 and 7.) This is also the point at which to discuss how the training will be complemented by professional development. A training programme must be developed to include a competency achievement planner and monitoring table setting out when and how the relevant competency levels will be achieved. For a graduate route 2 candidate supervisors and counsellors should make a benchmark assessment of the candidate in order to determine the levels of competency already achieved prior to starting the period of structured training.

Overall, it is important to give the candidate an insight into how you intend managing the partnership over the two years that will follow. Discuss and agree expectations – your own, and your candidate's. Make sure that all sides start with a clear vision and understanding of the events that will follow.

End the process by putting dates in your diaries for the necessary reviews, and make sure that these are agreed by all the parties required to attend. This ensures a proactive approach to managing the candidate's training, creates a good impression, and should ensure that the dates do not slip.

THREE-MONTHLY REVIEWS

The objective of the first three-monthly review is to allow the supervisor to verify that the candidate has got off to a good start with the training period. A lot of your views and opinions concerning the candidate's progress will come from your observations, discussions and contact with the candidate during the early part of the training period, which is itself a form of continuous assessment.

Care is needed at this early stage. All candidates develop at different rates, and their speed of development may vary from competency to competency. Sometimes candidates will make a slow start until they grasp or become familiar with basic concepts, but may then 'take off' with a rapid rate of development. The reverse may also apply, with candidates quick to pick up on initial points, but slower to transfer these into practice. Other candidates will follow a more consistent or 'straight line' rate of development. Remember to respect the diversity of approach – we are all different. Some candidates will also take a while to build confidence in their own ability and will appear slow initially, perhaps checking and double checking their work and what is required of them. It is important to keep a careful eye on the pace of development and to allow candidate's to develop their confidence and ability in their own time. Try not to compare one candidate's development with another.

The three-monthly reviews are also the point at which the supervisor concentrates on the candidate's management of the various APC documents. For the first three-monthly review, you will need to consider the following documents:

- the candidate's diary;

- the candidate's achievement record (log book)

- the candidate's professional development record; and

- the experience record.

Chapter 5 provides details of each of these documents. Ensure that the candidate is filling them in, or following them, as necessary.

You should use the APC guidance and the relevant pathway guidance when undertaking your assessment. Set aside a time and a place where you will be undisturbed for an hour or so. These reviews are very personal, privacy will be important.

The preparation of the experience record in advance of the review will help to focus the candidate's mind, and will allow you to open the meeting by asking how he or she feels the training is going and what progress has been made. This should ensure that your candidate gives you a very honest opinion and view.

Your discussion in the review will need to relate specifically to progress against the competencies. Ask the candidate to bring some examples of his or her work in order to either illustrate or remind you of the standard of work they have achieved. Try to 'stretch' the candidate's knowledge and experience. Ask what he or she has learned from a particular situation, what problems were faced, and how they were resolved, and how the experience could be transferred or adapted to deal with other areas of work. This ability to learn, and then transfer skills and experience to address other problems, will form an important aspect of the final assessment interview, particularly for those competencies which are required to level 3. It also allows you to check that the candidate has a good all-round view of any particular situation or technique, enabling them to repeat their actions under different circumstances.

At the end of each three-monthly review, you should firstly confirm any competency levels that you feel have been achieved and ask your candidate to update their candidate achievement record accordingly. You should keep a separate note of these achievements too. You should also complete the supervisor's progress report (an example form can be found at www.rics.org/apc). This asks you to make some notes concerning training to date, experience gained and ability of the candidate. At this first three-month stage your views concerning the ability of the candidate will be very general; however, you should keep an accurate record of your discussions around the competencies. Later reviews will be able to draw upon more information. The candidate will need to add some comments to the progress report too but these need only be brief. The notes on the report will also be useful for the counsellor, who will be involved in the six-monthly meeting and will need to review evidence of progress.

You should include on the progress reports details of the competency levels that you feel have been achieved at the time of the review. These will be formally recorded in the candidate's achievement record.

After filling in the progress report, it should be signed and dated by you and the candidate. Don't forget that when you do this, it is your professional integrity and reputation you are committing to paper. These progress reports are not submitted to RICS but are a very useful record of your candidate's training and development.

Two different examples of three-monthly progress reports are shown on pages 30 and 32.

Try to include a few 'action points' in the progress record as well, as this will signal to the candidate that he or she has an active partnership with you. Be realistic about what is achieved after the first three months – you may be able to sign off perhaps one or two level 1 technical competencies in the core and optional areas and one or two mandatory competencies to level 1.

At this stage you should also check the candidate's professional development record (template 3) to make sure that good progress is being made in this area. You would expect to see around 12 hours recorded here, although it could be more, if the candidate has undergone a lot of induction training in the first few months.

Note that the process carried out in the first three-monthly review must be repeated for subsequent reviews. For the later reviews, you will have much more information about the candidate and will be able to hold more focused discussions.

It is very important that a rigorous approach to the assessment of the competencies is established from the outset and retained throughout the course of the APC. Many candidates have expressed concern that their supervisors (and their counsellors) do not assess rigorously enough and this leads to a lack of confidence as the candidate is not sure whether they have truly met the levels of competence confirmed by their supervisors and counsellors. This can sometimes be taken with them right through the APC and into the final assessment and can have a seriously detrimental affect on their performance. It is crucial that competency achievements are 'hard won' although, of course, not unachievable!

SAMPLE SUPERVISOR'S PROGRESS REPORT A

To be completed every three months. (Note: This form is for the use of candidates and supervisors. It is not to be submitted to RICS.)

Date: 7 April 2009 (Period January 2009 to March 2009) – Observations on training to date, experience gained and ability of candidate:

Training to date

Laura has attended a two-day 'managing relationships at work' course. This course has added to her already developed ability to get on well with a variety of work colleagues and to appreciate their styles of approach.

Laura has built up a solid base in her first three months. In particular she has been able to produce financial appraisals and liaise with sub-consultants on such matters with the minimum of supervision.

- Report writing – soft market testing report, cash flow report

- Cash flow calculations – cash flow appraisal

- Brochure wording and questionnaire

- Evaluation process

- Market testing exercise with lead developers

- Involvement on report editing

Research – Laura has researched market evidence for residential space in connection with regeneration projects.

Client contact – Client exposure has increased dramatically and Laura has handled clients well.

Ability of candidate

Interpersonal skills – Laura is a good team member and fits in well. She has a good rapport with clients.

Written material – Laura has an appropriate use of business English and presents information in an organised manner.

Numerical – Laura has an excellent understanding of figures and the ability to use relevant systems and create cashflow spreadsheets.

Self-management – Laura makes good use of the working day to earn fees and has absorbed quickly the fundamentals of drafting and issuing invoices.

Competencies achieved at date of review:

- Level 1 – communication and negotiation; teamworking; valuation.

- Level 2 – development appraisal.

Signature: N Bursar   Date: 7 April 2009

Candidate's comments:

During my first three months within the Public Sector Consultancy team I have worked on a number of interesting large-scale local authority development schemes. The majority of the work carried out has been in the form of development appraisal analysis and other numerical tasks. I am thoroughly enjoying my work within the team and am continually learning from the other members of the team.

I feel a valued member of the department and believe that I have shown many of my strengths in the work that I have undertaken. During the next three months I hope to gain more experience in some of the areas in which I am not so competent, such as letter writing and report drafting. Also, during this period I would like to gain experience in general areas of surveying, such as landlord and tenant issues.

Signature: Laura Marshall   Date: 7 April 2009

SAMPLE SUPERVISOR'S PROGRESS REPORT B

To be completed every three months. (Note: This form is for the use of candidates and supervisors. It is not to be submitted to RICS.)

Date: 6 February 2009 – (Period January 2009 to March 2009) Observations on training to date, experience gained and ability of candidate:

Initially, Richard was given an overview of the firm and its organisation. This was followed by a review of the records kept (both paper and computer-based) and where to find them. The firm's intranet was also demonstrated, and details of where to find data relevant to Richard's day-to-day work provided.

Richard has attended a residential course, which covers the requirements of RICS guidance, legal aspects of his work and current environmental issues.

With regard to practical experience, Richard has so far mainly concentrated on residential property, although some commercial property has been covered. He has carried out valuations for inheritance and capital gains tax purposes, and is familiar with market value, the valuation of undivided shares and matters such as special purchasers and goodwill. He has encountered regulated, assured shorthold tenancies, and appreciates their impact on value.

Richard has also been introduced to the inspection of buildings. He has been able to assemble comparable evidence, analyse it and apply the results to valuations.

He has carried out reports for Housing Associations, and has negotiated values with solicitors and surveyors.

Richard has responded extremely well to the challenges he has faced, and is enthusiastic and keen to learn. He has worked hard and made good progress.

Signature: P Goody   Date: 6 February 2009

Competencies achieved at date of review:

- Level 1 – communication and negotiation; teamworking; valuation.

- Level 2 – development appraisal.

Candidate's comments:

I have learned a great deal in the short time I have been employed at the firm. Initially I was introduced to the bases of inheritance tax and capitals gains tax, and progressed from carrying out initial appraisals to inspecting and valuing residential properties under supervision.

The cases with which I have been involved so far have raised many issues in respect of statute and case law, which I am reviewing on a regular basis. I have begun to negotiate with agents, which has ensured that I am thorough with my valuations and has enhanced my oral communication skills.

In the next three months I hope to gain experience in valuing commercial properties, which I hope will raise additional landlord and tenant issues that will further my competence in this area.

I have gained much from my first three months at the firm, and have enjoyed being part of such a friendly and supportive team.

Signature: Richard Fitzwilliam   Date: 6 February 2009

SIX-MONTHLY REVIEWS

As far as the supervisor is concerned, the process for the six-monthly reviews will be very similar to that for the three-monthly reviews. You should carry out the actions noted above, and fill in another report on the appropriate template. The main difference at this juncture is the involvement of the counsellor. The counsellor is involved to assess the candidate against the competencies, to review overall progress, to provide a second opinion to that of the supervisor, and in so doing, to complete the counsellor's assessment and progress report. The idea is that the counsellor will take a strategic overview, to ensure that everything is on track. The counsellor should question progress, assist with areas of uncertainty, and most importantly, add an additional viewpoint to proceedings. This makes it clear why RICS prefers the two roles to be carried out by two separate people.

The counsellor should review the documentation, with a view to checking the number of days of training undertaken – it should be around 100 at the first six-monthly review, noted in the log book – and to check that progress is being made against the various competencies and levels. Do remember that the overriding approach, at the first six-monthly review, should be one that is progressive and reasonable for a candidate who is approaching the quarter-way mark: perhaps signed off to levels 1 or 2 in a number of competencies, with probably around three or four mandatory competencies completed. There should be a reasonable spread of days allocated to each competency in the log book. However, this will depend entirely upon the experience that is available and it is often the case that candidate's get little or no experience in some competencies in the early part of their APC.

In addition, check the number of hours recorded for professional development. At the first six-month stage you would expect to see around 24 hours recorded.

Having checked the documentation, the next step is to discuss the candidate's progress with the supervisor. There may be issues that require clarification, or over which you have concerns, or you may detect problems. After discussions with the supervisor, and if you feel that everything is on track, you should then speak to the candidate together with the supervisor. (If a joint meeting is difficult for reasons of distance, consider holding a telephone conference.)

If you have concerns and require a separate discussion with the candidate, then let the supervisor know that you are doing this. Make it clear that it is with the best intentions of both parties in mind and with the ultimate goal of ensuring that the candidate successfully completes the APC. After talking to the candidate, complete the loop by holding a further discussion with the supervisor, which may be merely to say that you are now happy, or perhaps to suggest some corrective action.

The precise approach adopted by the counsellor is likely to reflect the degree of contact the counsellor has with the candidate and supervisor. This may be regular and structured where all parties are working in the same office or in close geographical proximity, or much less frequent if the role is being carried out from a distance. In the latter case, a more formal approach to the six-monthly reviews may be required.

The counsellor should ensure that all actions and discussions are documented. You should state whether you and the supervisor had a joint meeting with the candidate or whether this was carried out separately. If you have adopted a tri-party approach, the completion of your report may be a repeat of the supervisor's report.

The counsellor should then confirm the competency levels that have been achieved and ensure that the candidate records this in the candidate achievement record. The achievements should be detailed in the progress report as well.

An example of a completed progress report is provided below.

A further six-monthly review is held every six months. This process should be repeated then, with more information being available to you about the candidate as the training period progresses.

SAMPLE COUNSELLOR'S PROGRESS REPORTS

To be completed every six months. (Please do not complete this section if you are also the candidate's supervisor.) (Note: This form is for the use of candidates and counsellors. It is not to be submitted to RICS.)

Date: 30 June 2009

Comments and recommendations:

On 29 June I held a joint meeting with the candidate and supervisor to discuss Richard's progress.

Richard has made an excellent start.

He has quickly become proficient at domestic valuations for national taxation purposes and has very capably dealt with residential valuations for other purposes. His valuations and reports are clear, well-considered and show a good understanding of the legal and valuation background.

He is a natural negotiator and is increasing in experience.

Richard is well-organised, conscientious, reliable, a quick learner and is already an invaluable and well-liked member of the team.

Over the final months with our team, we will arrange the compulsory purchase order experience that he now needs. I have changed Richard's work allocation, so that he will deal with a core West End location, in order to increase his commercial valuation experience.

Competencies achieved at date of review:

- Level 1 – communication and negotiation; teamworking; valuation.

- Level 2 – development appraisal.

Signature: A Sedley   Date: 30 June 2009

Candidate's comments:

In the last three months I have been given a greater variety of work, ranging from residential and commercial valuations, to the valuation of goodwill, compulsory purchase and compensation cases.

I carry out the majority of my inspections alone, although I discuss many of the issues that arise with my supervisor, to benefit from his experience.

In the next six months, I hope to improve my working knowledge of building construction and to gain further experience in compulsory purchase work and landlord and tenant issues, in order to attain the required competencies.

Signature: Richard Fitzwilliam   Date: 30 June 2009

ACTIONS AT 12 MONTHS (graduate route 1)

After 12 months it is important to have a strategic review of progress and to plan the final 12 months of training. This should involve the supervisor, the counsellor and the candidate and be undertaken alongside the relevant six-monthly review.

The experience record should be reviewed by the supervisor and counsellor, with a discussion based around any 'gaps' in experience that are revealed. Any such gaps will form the basis of the candidate's training needs over the next 12 months.

The supervisor and counsellor should then review the training programme, in particular the competency achievement planner and monitoring table to reflect any changes needed to the plans for the candidates training.

At this stage, you should also hope to see around 48 hours of professional development noted in the professional development template, linked to the candidate's training plan. Professional development activities should be allocated to one of three types of skills development:

- personal skills – linked to the mandatory competencies;

- technical skills – linked to the core and optional competencies; or

- professional practice skills development – linked to the professional practice competencies, code of conduct and conflicts of interest.

Generally it is expected that professional development is evenly weighted between these categories (approximately 16 hours per year for each type).

Again, it will be the gaps in professional development that will indicate exactly what is left to be achieved in the next 12 months.

The critical analysis

It is worth discussing with your candidate at this point the choice of project(s) for the critical analysis (a written report of a maximum of 3,000 words, comprising a detailed analysis of a project, or projects, with which the candidate has been extensively involved during the training period). There are now only around nine months before the candidate needs to write the critical analysis, and it is well worth starting to plan ahead. The candidate may wish to choose a topic that has already been covered in the first 12 months of training, or perhaps to identify a potential area of work or project that is coming on stream in the first six months of the second 12 months of training.

The choice of topic for the critical analysis is very important. Advice is given on this in the APC guidance and in the RICS publication *APC: your practical guide to success*. It is important to remember that the subject (or subjects) chosen do not have to be anything exciting or unusual. It is perfectly acceptable for a candidate to submit an analysis dealing with his or her role in the valuation of an ordinary small shop for a rent review (for example) – so long as it is a project in which he or she was extensively involved, and fulfils all the other requirements for the analysis, as discussed on page 68. Many candidates seem to feel that the only suitable choice of subject is something akin to the construction of a multi-million pound dam in the desert. Having said that, if your candidate has been involved in such a venture he or she is free to write about that.

Candidates should ensure that the subject(s) chosen will provide them with plenty of 'meat' for the analysis and their presentation on this. They should choose subjects they can 'get an angle on' – something on which they perhaps encountered difficulties, with a note on how these were overcome. Showy projects may look superficially impressive, but if the candidate has not been thoroughly involved, he or she will not be able to write or talk impressively on the subject. All in all, make sure your candidate gives him- or herself the best possible chance to demonstrate competence.

ACTIONS AT 18 MONTHS

At 18 months, with another supervisor's three-monthly and counsellor's six-monthly review underway, the most important point is to make sure your candidate has begun work on the critical analysis. The critical analysis is covered in more detail in chapter 5.

At this stage, you would expect your candidate to be about two-thirds to three-quarters of the way through the APC process, having completed about 300 days of training. He or she will probably be at about level 2 for most competencies (experience permitting!), with some perhaps attained to level 3. The candidate may well have completed about 72 hours of professional development.

ACTIONS AT 21 MONTHS

This is a critical three-monthly review as it may be the last before the candidate puts in their application for the final assessment. It is important to check your candidate's progress towards completion of the critical analysis and also to ensure that all other documentation is on course. Your candidate needs to be ready both in terms of paperwork and personal preparation for the final assessment, which will take place at some point over the next three to four months (deadline dates are available at www.rics.org/apc and are different for each pathway).

This may also be a useful point to start planning more detailed actions to prepare your candidate for the final assessment, such as presentation skills training and arranging any mock interviews.

ACTIONS AT 22 MONTHS

At 22 months, if your candidate has met all of the minimum competency levels and you (and they) feel that they are ready for final assessment then the candidate can apply for this if the opportunity exists to apply at this time. This will depend upon the exact time when the 22 months falls as the assessments are only held twice a year.

Both supervisor and counsellor must review whether overall the candidate is ready for final assessment. Have all the competencies needed been achieved? Has the candidate recorded at least 400 working days in the diary (or 200 for a graduate route 2 candidate) and have they indicated that they themselves feel ready? If so, you need to ensure that your candidate completes and returns the application form within the timescales set by RICS (as set out at www.rics.org/apc).

If, having reviewed your candidate's documentation, it is clear that they will not be ready for the final assessment then the application must be deferred. Entering an underprepared candidate for final assessment is likely to result in referral. It is a waste of everyone's time (and your money) to send a candidate with days of training incomplete, or a professional development record short on hours, and have them referred.

The responsibility therefore lies with the supervisor and counsellor to ensure that the candidate is competent in all of the required areas before applying for the final assessment. In essence, when you certify that the candidate is ready to sit his or her APC, it is your professional judgment that you are putting on the line. It should be remembered that 400 days/23 calendar months for a graduate route 1 candidate or 200 days/11 calendar months for a graduate route 2 candidate is a minimum training period. It is vital that candidates are not sent for final assessment too early. Failure (a referral) is not only disappointing and demoralising for them, but very time-consuming for everyone involved, including the supervisor and counsellor.

However, if you have carried out the actions noted in the rest of this book properly, then your candidate should be ready, and all of their documentation should be up to date.

Assuming the application is made there will then follow a period of approximately one month to complete and send to RICS the required documents for the final assessment presentation and interview.

ACTIONS AT 23 MONTHS

Supervisors and counsellors must ensure at this point that the candidate submits the correct documentation to RICS and that this is up to date. If the documentation is incorrect or incomplete RICS will return it and the candidate's final assessment date will be delayed by six months.

To ascertain your candidate's readiness for the final assessment and to make sure everything is ready to be submitted to RICS, you should review all documentation, to check that it is up to date and correctly filled in:

- Make sure the diary is up to date.

- Ensure the log book is complete and makes sense in terms of the number of days noted in total and reflects your opinion of the spread of training against each of the competencies.

- Ensure that all competencies have been reached and have been dated in the candidate achievement record (log book summary).

- Check the professional development record shows the appropriate number of hours of learning. For a candidate sitting the final assessment after two-year's training and experience it should note a minimum of 96 hours of professional development, with a balanced reflection of the requirements of the core, optional and mandatory competencies, and professional practice skills.

- Make sure the critical analysis has been completed.

- Check that the experience record is complete and accurately reflects the candidate's experience and supports the competency achievements recorded in the candidate achievement record.

The candidate must submit four bound copies of their final assessment submission including all the templates in their chosen pathway's Excel workbook. These are:

- Template 1 – candidate declaration – supervisors and counsellors will need to sign this template.

- Template 2 – candidate checklist.

- Template 3 – candidate achievement record – this must show compliance with the minimum number of recorded days needed and the achievement of the minimum competency requirements for the relevant pathway.

- Template 4 – professional development record – showing 48 hours of professional development for each year.

- Template 5 – education record.

- Template 6 – experience record.

- Critical analysis.

Once these submissions are received by RICS, the candidate will be sent confirmation of the date, time and venue of the final assessment (this will be one month prior to the assessment).

The assessment panel will receive all of the above documentation about four weeks before the final assessment interview. Remember, this will be their first impression of your candidate (and, incidentally, of you). Make sure it is a good one! It is important that the documentation is complete, well presented, signed and dated.

The final assessment will soon approach. If you are planning to organise mock interviews (or similar) for your candidate, or to set aside a time to provide advice on the final assessment, you will need to plan these actions well in advance. It may be around the 20 or 21 month reviews that you start planning any such activities.

A note regarding graduate route 3 candidates

Although this book generally provides advice on the supervision and counselling of candidates taking graduate routes 1 or 2 you may also be asked to act as a sponsor for a graduate route 3 candidate. As a sponsor you will need to know the candidate's work and will preferably be a colleague of the candidate. The candidate must have been accepted for the graduate route 3 to membership before you are needed to take any action as their sponsor.

Once they have been approved for graduate route 3, and assuming you are happy to act as sponsor, the candidate will ask you to verify that they have met the competency requirements for their chosen pathway. In order to do this you should undertake an interview with the candidate prior to which they should provide you with a list of the competencies they must achieve and the level required and a detailed list of how and when during their career they feel the competencies have been achieved. If you feel that the candidate has met the necessary competency requirements you should countersign their application.

Similarly for adaptation route candidates you may be asked to verify competence. Again a formal approach should be adopted for this, ideally by way of interview before confirming that the necessary competencies have been achieved.

Further guidance regarding graduate route 3 to membership can be obtained from www.rics.org/apc

Candidates for both graduate route 3 and the adaptation route will be expected to provide similar final assessment submissions as graduate route 1 and 2 candidates and to attend a final assessment interview.

PREPARATION FOR THE FINAL ASSESSMENT INTERVIEW

While you will play an important part in preparing the candidate for the final assessment interview, ultimately they will be on their own in the interview room. You should therefore encourage them to prepare themselves as much as possible. The RICS publication *APC: your practical guide to success* provides detailed guidance for candidates on the interview and how best to handle it. It tells the candidate exactly what to expect, and gives a wealth of useful tips on interview preparation.

Briefly, the interview will comprise a welcome from the panel of assessors, a ten-minute presentation by the candidate, based on the critical analysis that has been submitted, and a period of questioning, on both the presentation and analysis, and on wider issues relating to the candidate's experience and training.

From the point of view of the supervisor or counsellor, how can you help your candidate prepare for the final assessment interview? Really the best form of preparation is practice! Encourage your candidate to give a presentation to you, based on their analysis, and to undergo some sample questioning.

With the presentation, check that this does last the required ten minutes – the effects of adrenalin can often make people talk too fast, or cover ground more quickly than they would in a more relaxed setting. Note that with only ten minutes in which to present the candidate will not simply be reiterating the contents of the critical analysis, but will be giving the panel an overview of it. This could include discussing the reasons for choosing that particular subject, talking about wider issues relating to it, and extrapolating lessons learnt from it.

Candidates should not use a laptop for their presentation as projection facilities are not available at assessment centres and after all it is the candidate's personal communication and presentation skills that are being assessed. You should consider the choice of presentation aids made by your candidate and consider whether you feel these to be appropriate.

Ask the candidate to consider different approaches that could be adopted in a similar situation. You may wish to encourage your candidate to have three or four extra points in mind when giving the presentation, which expand the matter contained in the analysis itself, and give them something extra to talk about. It is vital that the candidate can take a 360 degree view of the project that is the subject of the analysis – the panel may ask questions about it from an unexpected angle. This of course is why it is vital that the candidate has been heavily and thoroughly involved with that particular project.

Moving on from the questioning based around the critical analysis, ask the candidate the kind of questions based on their experience that you imagine the assessment panel would put to them. Remember that the final assessment interview will adopt a 'competency-based' approach (see page 23 for more information on this). In brief, this is an approach based on practical experience. It does not set out simply to test what a candidate knows, but to assess how they can put this knowledge into practice. In essence, it seeks to ask, 'Is the candidate competent in this area?'

To assess practical competence, the assessors are likely to ask questions such as: 'Give me an example of a time when...', or, 'I see you have been involved in .... Tell me how you went about this.' This approach aims to allow the candidate to demonstrate his or her skills and abilities across the range of competencies covered during the training period.

With those competencies that are required to be achieved to a level higher than level 1, the assessors will aim to move beyond the candidate's knowledge and experience, to test their more general attitudes and behaviours. Try to adopt a similar approach when questioning your candidate.

If you feel you are too intimately involved with the candidate's progress to carry out this questioning objectively, consider asking a colleague to perform this function for you – or at the very least to sit in the room with you while you hold the practice interview, and compare notes later.

A big problem with any interview – and especially one with so much resting on it – can be nerves. There are various techniques for countering these, which are recommended to the candidate in *APC: your practical guide to success*. Supervisors and counsellors can also help by giving the candidate a chance to practise under interview conditions. Hold a practice interview in a formal setting, in formal dress, with a colleague who is less well-known to the candidate in attendance. Impress on the candidate that if they can survive and 'pass' this experience, then they can succeed in the actual interview. Try not to make this a more difficult experience than the final assessment itself as this can undermine a candidate's confidence.

During the practice interview, ask yourself the following questions:

- Does the candidate's presentation truly reflect the content of the critical analysis, and is his or her role in the project chosen clearly articulated and accurate?

- Do you have a proper sense of what the candidate has achieved over the training period?

- Can the candidate answer all of your questions properly – or is he or she under-prepared, or answering by rote, in a particular area?

- Would you be happy, if you were an assessor, to admit this candidate to membership of RICS?

If the answer to any of these questions is no, then consider exactly what the candidate must do to rectify the situation.

Be ready to give honest feedback to the candidate after holding practice interviews – even on matters as personal as formality of dress, use of eye-contact and body language, speed of speech, your level of interest in the presentation, and their levels of nerves and enthusiasm. Honesty is by far the best policy at this stage in the proceedings. If you do have to make a honest criticism, make sure it is constructive (i.e. tell the candidate what they can do to put the problem right), and temper it with praise in other areas. A confident candidate is much more likely to be successful than a demoralised candidate. If you've got to this stage in the APC process, then you know they can do it – make sure they know that too.

Finally, do the obvious. Encourage your candidate to have a restful week or so before the interview. Make sure they know where the assessment centre is, and how they will get there.

If you have carefully followed the advice in the rest of this book, your candidate should have nothing to worry about. The final assessment will be simply the formal end to a process of continuous and progressive assessment: the 'stamp of approval' on what has gone before. There is therefore no need for your candidate to feel that everything is riding on this interview, and to build it up into an insurmountable obstacle. By the time of the final interview, it is in fact too late for the candidate to 'save the day' by pulling off a fine performance.

Figures 2, 3 and 4 below and overleaf show the APC assessment panel mark sheet for the three components of the final assessment.

Critical Analysis Title	Notes
Suitable project(s)/ process selected for the critical analysis.	☐ Met ☐ Not Met
Key issues identified.	☐ Met ☐ Not Met
All relevant options have been considered and reasons given for those options which have been rejected.	☐ Met ☐ Not Met
Clear explanation of and reasons given supporting proposed solution.	☐ Met ☐ Not Met
Conclusion and critical appraisal of the proposed solution and outcomes.	☐ Met ☐ Not Met

Analysis of experience gained, demonstrating the candidate's learning and development.	☐ Met ☐ Not Met
A good display of professional and technical knowledge and problem solving abilities.	☐ Met ☐ Not Met
Overall standard of: • written presentation • layout • spelling • grammar • graphics	☐ Met ☐ Not Met

Figure 2 Mark sheet – critical analysis

Oral communication skills	☐ Met ☐ Not Met
Clarity of thought and structure	☐ Met ☐ Not Met
Presentation skills: • eye contact • body language • voice projection • visual aids (if any)	☐ Met ☐ Not Met

Figure 3 Mark sheet – presentation

Attach the Achievement Record – Log Book Summary pages ONLY (Template 3).
Relate your marking to the competencies noted in Template 3.

	Notes
Mandatory competencies	☐ Met ☐ Not Met
Technical Core competencies	☐ Met ☐ Not Met
Technical Optional competencies	☐ Met ☐ Not Met
Rules of Conduct/Ethics	☐ Met ☐ Not Met
Professional development	☐ Met ☐ Not Met

Figure 4 Mark sheet – interview

AFTER THE FINAL ASSESSMENT

Sometimes, things can of course go wrong, and candidates are referred. A referral allows them to retake the final assessment six months (or more) afterwards. Referrals are discussed in detail in chapter 8.

BEYOND SUCCESS

Successful candidates are a credit not just to themselves, but to their supervisors and counsellors, and to their firms. Make sure they know that you are proud of their achievement. Encourage them to feel that they have not just passed a test but have taken the first steps into an exciting and fascinating profession. The rest of their career begins here – and hopefully much of it will be spent with the firm that helped them to initial success.

A successful candidate will be a good employee and an advertisement for other good employees. The investment put in by the employer, supervisor and counsellor throughout the APC process will now start to pay off – for all concerned.

A last word of advice

For you to have the confidence that you can guide your candidate correctly, you need to know exactly what you should be doing at each stage in the process. This book and the official APC guidance should help you to ensure that the process is completely clear in your mind.

5 Documentation for the APC

It is essential that supervisors and counsellors familiarise themselves with all the documents required for the APC, so that they can understand exactly what is required of them and the candidate.

This chapter will introduce you to the following documents:

- enrolment form;
- change of employer form;
- structured training agreement;
- professional development record;
- diary;
- candidate achievement record (including the log book);
- experience record; and
- critical analysis.

Below, we consider what each document is, and what function it performs. There is then a practical guidance section for supervisors and counsellors, explaining their role with regard to each document.

ENROLMENT FORM

An enrolment form is obtainable from the RICS Contact Centre (see page 125).

You and your candidate MUST read the APC guidance carefully. Before the APC process can start you must send the completed enrolment form with the correct fee to the RICS Membership Operations department. Details of the fees are included with the enrolment form. You must submit all the required material otherwise the enrolment cannot be accepted. Once accepted, RICS will confirm the registration and give the candidate a start date for recording their experience.

A candidate cannot enrol until he or she is in employment – but it is in your interests to get the ball rolling soon after that , so that you have a fully trained and qualified member of staff as soon as possible.

Before your candidate can reach the final assessment, a minimum of 400 days of experience, within (at least) 23 calendar months, must have been completed for graduate route 1 or 200 days in 11 calendar months for graduate route 2 candidates. With final assessment interviews only held twice a year (details on www.rics.org), a delay of just a few weeks in enrolling could put the final assessment back by six months.

It is important to take note of the APC guidance and pathway guidance, also available at www.rics.org/apc

All the templates needed for the APC are included within an Excel workbook for the candidate's pathway, available at www.rics.org/apc

Practical guidance

To show the candidate your commitment from the outset, offer to agree a time and date to discuss the application form. You can then provide any necessary support and advice on this, if required. Stress the importance, as noted above, of returning the application documentation promptly.

CHANGE OF EMPLOYER FORM

It is mandatory to keep RICS informed of changes of employer. Candidate's can email contactrics@rics.org with the new details or a 'change of employer' form is available at www.rics.org/apc. This needs to be completed and returned to RICS should the candidate leave your firm's employment during the training period.

Practical guidance

If your firm takes on a new employee who has already enrolled on the APC with another employer, you should check that the appropriate notification has been sent to RICS, giving details of the new employment. Failure to do so may delay the candidate's final assessment, as RICS may not recognise the 'gap' as counting towards the minimum training period.

You may also wish to remind any candidates leaving your employment that this form must be completed and returned to RICS, with details of the new employer.

STRUCTURED TRAINING AGREEMENT

Structured training is, as the name suggests, a structured approach to the delivery of training and experience over any given period. All firms registering new APC candidates are obliged to have a structured training agreement (STA) in place.

The STA is a written document that formalises the organisation's intended policy in implementing the RICS requirements for the APC. The document must be approved by an RICS training adviser.

A useful template for the development of a STA can be downloaded from www.rics.org. However, this is not a mandatory template and organisations may formalise their STAs in other ways so long as these demonstrate the ways by which the requirements of the APC will be met. STAs can be developed and approved at a national level (for firms with offices throughout the UK) or at a regional level (just for one or a few offices in one region). The relevant approval will be noted on the RICS website in the list of firms with approved STAs. The STA is developed for organisations rather than individual candidates and once approved is tailored and used for all candidates (or all candidates within the region for a regional STA). The STA must be used in conjunction with the official APC guidance. It **does not** replace them. Firms that have taken on several APC candidates in the past may already have an agreement in place (although this must be tailored to meet the individual requirements of each candidate).

The STA template provides guidance only – you will need to tailor the agreement to your organisation. The following information should generally be included within a STA:

● statements of commitment to the APC and to the delivery of a structured training programme;

● information on the organisation and areas of activity;

● information on the organisation's policies regarding training and the APC, including how the professional development requirement will be met and supported; and

● a commitment to setting out a training programme for each candidate providing arrangements and timescales for monitoring the training programme.

The training programme devised for each candidate should include:

● a competency achievement planner setting out the chosen competencies and the timescales within which the supervisor and counsellor expect the candidate to achieve each level of each competency; and

- a monitoring table (or similar) setting out the experience that will be available for each level of each competency that the candidate needs to achieve.

Practical guidance

For employers who do not have a training agreement in place, the template and guidance will provide a useful discussion document for the supervisor and candidate to plan and agree the training for the next two years. Supervisors and counsellors should also approach an RICS training adviser (RTA) for additional support on drawing up the STA. (For more information on RTAs, see chapter 9.)

Employers who have recently trained APC candidates will already have a training agreement in place and will be familiar with RICS requirements. All that will be needed is a degree of tailoring to each individual candidate and the development of a specific training programme in relation to the timescales and competencies. RTAs can advise on both how to set up a new STA and how to set out a training programme for a new candidate.

The STA is a demonstration of your commitment to APC training and may be used when advertising for graduates. It is, if you like, the RICS training advisers' kite-mark of excellence. Additionally, of course, having an STA in place saves time when considering the training plan for each new candidate.

There sometimes seems to be a view that larger employers stand a better chance of getting candidates through the APC, as they may have more opportunities for the candidate to gain a breadth of experience. However, in fact, everything depends on the firm itself. If an employer understands what needs to be done, and takes time to tailor requirements to the candidate's needs, then size is of no issue. A smaller employer, where the candidate is in closer proximity to all other members of staff, and to the decisions being made, may have more scope for flexible development than a candidate in a larger, more impersonal firm, where the training programme may be more rigid.

Whatever the size of the practice, supervisors and counsellors must take a very personal and close interest in the candidate's progress. Whether this happens in a large or small firm, that approach is the best possible training ground.

When the STA has been drawn up, it must be sent for approval by an RICS training advisor. It is not sent to RICS. Once the RTA approves the documents he or she will notify RICS and confirmation of the approval will be sent.

PROFESSIONAL DEVELOPMENT RECORD

For each 12 months of practical training that is completed, the candidate must also undertake an annual minimum of 48 hours of professional development. The only exception to this is when a candidate is on a part time or distance learning accredited degree when the final year of their course can count towards the professional development for that year. Ideally though, this should be supplemented by other professional development activities. The idea behind professional development is that it provides an opportunity for the candidate to acquire some of the additional skills and knowledge that it is not always possible for the employer to provide within the week-to-week business of their practice. This could apply to any of the competencies.

Professional development should be designed to complement and support the candidate's on-the-job training and development. It may comprise formal training courses or more informal types of learning, such as structured reading, distance learning programmes and secondments. However, it should be noted that while structured reading and private study are acceptable they should not form the entirety of professional development in any 12 months of structured training. A mix of modes of learning is important. While it is the candidate's ultimate responsibility to plan and acquire professional development, it is important that the supervisor and counsellor take an active interest in this and assist with the evaluation.

Like every other aspect of the APC, professional development should be planned and structured, with the aim of completing the requisite number of hours. The 48 hours should reflect the requirements of the candidate's chosen pathway. Around 16 hours per year should be allocated for technical skills development linked to the core and optional competencies of the pathway. A further 16 hours should be allocated for professional practice skills development, linked to those competencies associated with professional practice, code of ethics and conflicts of interest; in addition, 16 hours should be dedicated to skills development linked to the requirements of the other mandatory competencies. (Advice on how to meet some of these requirements is provided in chapter 7, where the mandatory competencies are discussed in more detail.) The table overleaf represents the recommended annual breakdown of professional development.

Type of development	Related competencies	Hours
Technical skills development	Core and optional competencies	16
Personal skills development	Mandatory competencies	16
Professional practice skills development	Competencies associated with professional practice, code of ethics and conflicts of interest	16
Annual professional development hours		48

Having decided on the plan, the hours of learning achieved by the candidate are then recorded in the professional development record over the course of the APC training period. This record should also include the candidate's view of the outcome of each activity in terms of their learning. This record should be checked by the supervisor and counsellor regularly, to ensure that good progress is being made.

Practical guidance

You should ensure that your candidate understands what counts as professional development.

Professional development is learning that is relevant to the candidate's professional role and learning needs. The most important aspect is the learning outcome. Many activities can qualify, not just formal training courses. The candidate's record of professional development should ideally include a balanced mix of formal courses, structured reading and other activities.

The candidate needs to ensure, when choosing a learning activity, that some significant learning occurs.

Remember also that the activities chosen should complement the requirements of the candidate's main area of work. The panel at the final assessment interview will check carefully that there has been a sensible, structured approach to professional development over the training period. There should be a clearly defined relationship between the topics selected for professional development purposes and the competencies. The training suggested for the various mandatory competencies in chapter 7 may be useful for candidates planning their professional development. With 32 of the 48 hours designed to reflect the mandatory competencies, this may be a useful place to start.

Bear in mind that the 48 hours per year recorded for professional development are over and above the 400 days (200 days for graduate route 2) to be recorded for the competencies. Although the professional development hours must reflect the requirements of the competencies, the candidate cannot 'double-count' experience. He or she must decide where to record a particular activity – under a competency heading in the log book or in the professional development record. This should not pose a problem: over the course of 24 months, there is more than enough time to gain the necessary experience in all areas. Remember also that the diary only records experience against the core and optional competencies so where a learning activity relates specifically to a mandatory competency this is often a good opportunity for the candidate to record professional development.

If, in discussion with the candidate, you feel that there is a need for variation regarding the number of hours allocated in the typical annual plan shown above, you must ensure that an explanation of this departure from the norm is included in the professional development record. There are many good reasons for such a departure – candidates who are highly experienced in a particular area of work, for example, may spend less time on professional development relating to the core competencies, and more on those relating to the more unfamiliar mandatory competencies.

In the context of life-long learning, the professional development undertaken as a component of the APC is a precursor to the candidate's continuing professional development (CPD) commitments upon qualification. Professional development is essential for all surveyors to continue to grow and learn.

There is more information on the professional development record in chapter 4.

DIARY

Candidates are obliged to keep a record in a diary to show how their day-to-day training and experience is being built up. This diary should ideally be kept electronically. A diary template may be found on www.rics.org

The detail contained in the diary will be used for three specific reasons:

- to complete the log book;

- to help with the assessment of competencies by the supervisor and counsellor; and

- to assist with the completion of the experience record.

The diary can also be requested by an RTA at any time (and can also be requested by the APC final assessment panel) – so it is important it is always up to date.

Practical guidance

Candidates record their experience in the diary with reference to their chosen competencies. Entries should be recorded in periods of no less than half a day. The level of detail needed should be sufficient to allow the candidate to summarise the diary activities in the experience record. A statement such as 'undertook an inspection of a commercial property' is unlikely to be particularly helpful in preparing the experience record. However, an entry noting: 'Inspected a shop at [insert street, town] for rent review purposes. Large, double-fronted shop in secondary location, with good service access and parking to rear' will provide the candidate with a memory-jogger for those reports and for preparation for the final assessment.

CANDIDATE ACHIEVEMENT RECORD (including log book)

The candidate achievement record includes a log book and a log book summary. The log book must be completed by the candidate every month. The log book is a summary of the candidate's diary, showing the training – in number of days – grouped under the competency headings.

The log book summary provides details of the number of days training each year, again grouped under the competency headings. Importantly this form also includes the dates of the candidate's achievements of the various competency levels as assessed by the supervisor and the counsellor.

The candidate achievement record will be sent to RICS as part of the final assessment submission, to be used by the assessment panel. The information it contains is very useful to the assessors, as it provides an immediate snapshot of the candidate's competency achievements and weightings of work experience. In conjunction with the experience record it will be used to structure the final assessment interview to obtain the correct balance of questioning relative to the candidate's experience.

Practical guidance

It is vital that the log book provides an accurate reflection of the balance of a candidate's experience in the relevant competency areas. There are no specified numbers of days required in each competency and the amount of experience needed for a candidate to achieve a given level of competence will vary from candidate to candidate. This will depend upon whether the candidate has any previous experience and, of course, how quickly they pick up the skills necessary. Although there is an overall time requirement the APC is not about time served but about achieving the necessary skills and competence.

For graduate route 2 candidates previous experience may mean that they reach the required level of one competency much quicker than another and, in fact, some of the competency levels will have been reached prior to starting the period of structured training.

As set out previously the minimum training requirement is for 400 days over 23 calendar months for graduate route 1 or 200 days over 11 months for graduate route 2 candidates. Bear in mind that mandatory competencies are not recorded as part of the 400/200 days, but candidates must explain the experience they have gained in these competencies in the experience record and they must also be assessed by the supervisor and counsellor.

In any 12 months a typical candidate might work for around 45–46 weeks, or 220–230 days. So in theory, the requirement of 400 days can quite easily be met.

The training and experience needed to achieve the mandatory competencies will be acquired as part of the training and experience for the core and optional competencies. For example, health and safety, customercare and communication skills are all likely to be addressed when carrying out training for almost any core or optional competency.

There will also, of course, at times be an overlap between the core and optional competencies. Training for one will be gained in the course of training for another. A commonsense approach is needed with regard to apportioning the time spent on each competency in the diary.

Note that professional development hours are recorded in a separate record and not in the log book. At times it may be necessary for the candidate to make a decision as to where exactly to allocate days and hours of experience, if there is an overlap in the requirements.

Overall, each individual – the candidate, the supervisor and the counsellor – must appreciate that training and experience rarely breaks down neatly into categories. Candidates will be learning all the time – it is just a case of deciding where and how best to record the experience. Experience in one area helps to build experience in another. Even if it cannot be recorded in two categories, it is never wasted.

One final piece of advice is not to attempt to cover too many competencies over the training period, thus spreading the candidate's experience thinly over too many areas.

The candidate achievement record is also used to record the competencies that the candidate achieves and the dates of these achievements. The supervisors and counsellors progress reports discussed in chapter 4 should be used to provide a formal record of the review and to underpin the competency achievements recorded in the achievement record.

A discussion should take place at each of the three- and six-monthly reviews around each of the competencies, and the supervisor and counsellor can then take a view as to whether any of the competency levels have been achieved. Supervisors and counsellors must advise candidates of these achievements after each of the three- and six-monthly reviews and ensure that the candidate updates the candidate achievement record accordingly. An example of the candidate achievement record for the commercial property pathway is shown in figure 5 opposite.

EXPERIENCE RECORD

The experience record is a record of the candidate's experience and training relevant to each of the levels of each of the competencies. Candidates should complete this record before each of the supervisor and counsellor assessments. The experience record should, in effect, provide a summary of their diary for each of the competencies where they have gained further experience in that period. In total, for each level of each competence achieved they are asked to aim to have written up to 200 words. Candidates should use this record to clearly demonstrate the projects and work with which they have been involved and to show how these are linked to the relevant competencies.

Practical guidance

You should aim to make the experience record the focal point of your meetings with your candidate at the three- and six-monthly intervals.

RICS

Candidate Name - 1234567
Template 3
Select a Route

Candidate Achievement Record

Commercial Property

Log Book Summary

Completed by the Candidate, Supervisor and Counsellor
The competencies listed below are those that the candidate believes they are competent in and wishes to be assessed against and relate directly to the minimum requirements as stated in the APC/ATC Requirements and Competencies Guide (2006).

Number and Competency Title		Dates Achieved		
		Supervisor	Counsellor	
	Level	dd/mm/yyyy	dd/mm/yyyy	Assessor Use
Mandatory Competencies				Met
M001 Accounting principles and procedures	1	dd/mm/yyyy	dd/mm/yyyy	
M002 Business planning	1	dd/mm/yyyy	dd/mm/yyyy	
M003 Client care	1	dd/mm/yyyy	dd/mm/yyyy	
	2	dd/mm/yyyy	dd/mm/yyyy	
M004 Communication and negotiation	1	dd/mm/yyyy	dd/mm/yyyy	
	2	dd/mm/yyyy	dd/mm/yyyy	
M005 Conduct rules, ethics and professional practice	1	dd/mm/yyyy	dd/mm/yyyy	
	2	dd/mm/yyyy	dd/mm/yyyy	
	3	dd/mm/yyyy	dd/mm/yyyy	
M006 Conflict avoidance, management and dispute resolution proce	1	dd/mm/yyyy	dd/mm/yyyy	
M007 Data management	1	dd/mm/yyyy	dd/mm/yyyy	
M008 Health and safety	1	dd/mm/yyyy	dd/mm/yyyy	
	2	dd/mm/yyyy	dd/mm/yyyy	
M009 Sustainability	1	dd/mm/yyyy	dd/mm/yyyy	
M010 Teamworking	1	dd/mm/yyyy	dd/mm/yyyy	

Candidate Name - 1234567
Template 3

Log Book Summary (cont)   **Commercial Property**

Number and Competency Title	Level	Dates Achieved		Year		Total Days	
		Supervisor dd/mm/yyyy	Counsellor dd/mm/yyyy	1	2		Assessor Use
Technical - Core Competencies							Met
T044 Inspection	1	dd/mm/yyyy	dd/mm/yyyy				
	2	dd/mm/yyyy	dd/mm/yyyy				
	3	dd/mm/yyyy	dd/mm/yyyy				
T057 Measurement of land and property	1	dd/mm/yyyy	dd/mm/yyyy				
	2	dd/mm/yyyy	dd/mm/yyyy				
	3	dd/mm/yyyy	dd/mm/yyyy				
T083 Valuation	1	dd/mm/yyyy	dd/mm/yyyy				
	2	dd/mm/yyyy	dd/mm/yyyy				
	3	dd/mm/yyyy	dd/mm/yyyy				
		Total					

Candidate Name - 1234567
Template 3

Log Book Summary (cont)   **Commercial Property**

Number and Competency Title	Level	Dates Achieved		Year		Total Days	
		Supervisor dd/mm/yyyy	Counsellor dd/mm/yyyy	1	2		Assessor Use
Technical - Optional Competencies							Met
Select From Dropdown List	1	dd/mm/yyyy	dd/mm/yyyy				
	2	dd/mm/yyyy	dd/mm/yyyy				
	3	dd/mm/yyyy	dd/mm/yyyy				
Select From Dropdown List	1	dd/mm/yyyy	dd/mm/yyyy				
	2	dd/mm/yyyy	dd/mm/yyyy				
	3	dd/mm/yyyy	dd/mm/yyyy				
Select From Dropdown List	1	dd/mm/yyyy	dd/mm/yyyy				
	2	dd/mm/yyyy	dd/mm/yyyy				
	3	dd/mm/yyyy	dd/mm/yyyy				
		Total					
Technical - Optional Plus Competencies							Met
Select From Dropdown List	1	dd/mm/yyyy	dd/mm/yyyy				
	2	dd/mm/yyyy	dd/mm/yyyy				
	3	dd/mm/yyyy	dd/mm/yyyy				
Select From Dropdown List	1	dd/mm/yyyy	dd/mm/yyyy				
	2	dd/mm/yyyy	dd/mm/yyyy				
	3	dd/mm/yyyy	dd/mm/yyyy				
		Total					

Figure 5 Candidate achievement record

Candidates need to be proactive in managing their progress and should, therefore, prepare this record in advance of these meetings, in order that it can be used by you in assessing their progress. The experience record provides a useful summary of the experience that the candidate has obtained within each competency, which is not always as clear in the diary, which is in chronological order rather than in order of competencies.

It is important that you provide feedback to the candidate on the information they have provided in their experience record. When you confirm achievement of a level of competence you should ensure that the information the candidate has presented in their experience record reflects the level of competence achieved. The pathway guidance provides a useful point of reference for this as the experience outlined in the experience record should align with that suggested for the relevant pathway.

CRITICAL ANALYSIS

The critical analysis is a written report of a maximum of 3,000 words, comprising a detailed analysis of a project, or projects, with which the candidate has been extensively involved over the training period. Chapter 4 discusses in detail the choice of subject for the analysis.

The aim of the analysis is to give the assessors practical evidence of the candidate's ability to work effectively within the competency requirements and to show critical reflection. The analysis should indicate a thorough understanding of the project concerned and the processes that were followed. In the final assessment, the candidate will be questioned on the approach outlined in the analysis, as well as on wider issues surrounding the report. It is therefore vital that the candidate has a good all-round view of the project.

The conclusion of the analysis should include a critical appraisal of the project(s), together with some reflective analysis of the lessons learnt and experience gained.

The main failing of critical analyses that do not satisfy the assessors is a lack of early preparation. The choice of subject is either too simplistic or is not a reflection of what the candidate has been doing during the training period. Remember that the analysis must be a reflection of the candidate's own work, and of something in which they have been directly and heavily involved.

Another common failing is omitting to follow the APC guidance in terms of the format of the report.

To summarise this, the analysis must be word-processed and must be a maximum of 3,000 words (not including the appendices). It is advisable to include photographs and plans where these are relevant, which must be no larger than A4 size when folded. It must be signed and dated by the candidate, and certified by the supervisor and counsellor. The standard of presentation should be as high as for any other report leaving your office and representing your organisation.

Official guidance calls for the critical analysis to fall under the following headings:

- key issues;

- options (to include reasons for rejecting options that may not be feasible);

- the proposed solution; and

- conclusion and analysis of the experience gained.

It is vital that all these headings exist, and that the commentary under them is relevant to each one. Too often, the analysis is written in the style of a report to a client, and becomes too much of a 'diary of events':

lateral thought is missing, there is no critical appraisal or reflective analysis, and the report therefore fails to meet the required standards.

Supervisors and counsellors should also make sure that the analysis allows the candidate to demonstrate ability across a number of competencies. To do so, it is not necessary to formally list each competency and discuss how this has been covered, but simply to step back and consider what exactly it is the candidate has done, and how this fits with the requirements of the competencies. If this is clear to you, it should be clear to the final assessment panel too.

There is further advice available for candidates on the critical analysis in the APC guidance, as well as in the RICS publication, *APC: your practical guide to success.*

Once the analysis is written, the best help that a supervisor and counsellor can give a candidate is to stand in the shoes of an assessor. Ask the candidate to provide you with a first draft and check that the key issues referred to above are covered. Consider the layout, presentation, spelling and grammar – are these of a high standard? Stand back and analyse the technical and professional impact: are you satisfied that it is truly representative of your candidate's work, and that it meets with the required level of the appropriate competencies? Always check the candidate's final draft very carefully both as a proofreader but also for technical accuracy and to ensure that the candidate has followed the guidance. Where candidates have included copies of any standard material, for example the organisation's terms of engagement, do ensure that these also follow RICS guidance! One example of this was where a candidate included their organisation's terms of engagement but these made no reference to the complaints handling procedure!

The candidate will be required, at the final assessment interview, to give a ten-minute presentation on the analysis.

6 The competencies

WHAT ARE THE COMPETENCIES?

The APC is a test which ensures that trainee surveyors are competent in terms of the standards set by RICS. To be 'competent' is to have the skill or knowledge to carry out a task or function successfully – this ability can vary from being merely able, to being expert in a particular sphere of activity.

A 'competency' is a statement of the skills or abilities required to perform a specific task or function. It is based upon attitudes and behaviours, as well as skills and knowledge. The training structure of the APC requires candidates to achieve a certain set of competencies. These are a mix of technical and professional practice, interpersonal, financial, business and management skills.

The competencies a candidate undertakes depend on the pathway to membership being followed. The pathways themselves are linked to the respective RICS professional group. The APC guidance sets out the required competencies and levels of attainment for each APC pathway – candidates, supervisors and counsellors must all read this guidance carefully.

In a broader context, the use of competency-based assessment is becoming a global approach to assessment both in academic environments and within the occupations and professions. It is felt that such an approach results in candidates and employees who have not just theoretical knowledge, but the ability to put this into practice. The basic philosophy behind competency-based assessment is that evidence of past performance and behaviour is the best prediction of future performance and behaviour.

LIST AND LEVELS OF COMPETENCIES

The APC guidance sets out what the candidate needs to achieve by way of skills and abilities over the training period. It notes the requirements of each APC pathway, with a list of the number of competencies to be covered during the training period and the level of attainment required in each. The guidance also includes the full list of competencies in alphabetical order, giving each a reference number for use on the achievement record and spelling out precisely what each entails.

The competencies cover three levels of attainment, which are progressive in terms of skills and abilities:

- Level 1 – knowledge and understanding of the areas covered by the competency;

- Level 2 – application of this knowledge and understanding in practical situations; and

- Level 3 – ability to provide reasoned advice and depth of technical knowledge.

Depth of technical knowledge referred to in level 3 must be viewed in the context of the level of experience a candidate can reasonably be expected to achieve after just two years (for a graduate route 1 candidate). Candidates will reach each level in a progressive and logical order – as the above should indicate, it is not possible to move from level 1 to 3 without passing through level 2.

Consider the example below of the building surveying competency of 'building pathology', which progresses in complexity across the levels.

Building pathology

Level 1: demonstrate your knowledge and understanding of building defects including collection of information, measurements and tests.

Level 2: apply your knowledge to undertake surveys, use survey and other information to diagnose cause and mechanisms of failure.

Level 3: provide evidence of reasoned advice and appropriate recommendations, including the preparation and presentation of reports.

A typical final assessment question for this competency might be:

'I notice from your summaries of experience that you have carried out building surveys on a number of traditionally built 1960s houses. Describe how you went about those surveys [level 1]. What were the common types of failure to brickwork observed while carrying out those inspections? What were the causes and how did you diagnose these? [level 2]. What method of repair did you recommend to your client, and why? What sort of issues did you include in your final report and recommendations? [level 3].'

The levels of competencies are discussed in more detail below.

TYPES OF COMPETENCIES

The APC has three types of competencies: mandatory, core and optional. The core and optional competencies are referred to as 'technical' competencies.

The mandatory competencies are considered by RICS to underpin all practice as a chartered surveyor. All candidates on all pathways must achieve these.

The core competencies are the primary skills of the candidate's chosen pathway. For example, 'valuation' is a core competency in the valuation pathway. Similarly, 'contract practices' is a core competency in the quantity surveying and construction pathway.

The optional competencies are selected by the candidate as additional requirements for the chosen pathway. For each pathway, the candidate must achieve all of the core competencies, and a certain number of optional competencies. The optional competencies are not so called because they are optional altogether – the candidate has a choice of which ones to achieve.

You will note from the APC guidance that with most pathways candidates will start by selecting optional competencies from a closed list, and then there is a further requirement based on the full list of technical competencies including any not already chosen from the 'closed' list. It is important that the guidance for each pathway is considered carefully so as to ensure the correct selection of competencies.

HOW DO YOU CHOOSE THE COMPETENCIES?

First of all, remember that all candidates must achieve the mandatory competencies – there is no choice in that.

The other competencies are chosen by reference to the candidate's APC pathway. On enrolment, the supervisor, counsellor and candidate select the most appropriate pathway (all pathways relate to specific RICS professional groups).

The pathway guidance, available at www.rics.org/apc, provides further advice regarding the experience requirements within each of the competencies. Set out below is an explanation of the experience required in the building pathology competency taken from the building surveying pathway and gives further guidance on the areas of experience that a candidate will be expected to cover.

Building pathology

Level 1: Demonstrate knowledge and understanding of:

- typical defects relating to typical buildings found in your locality that you may have come across and explain cause and effect of these;

- building defects likely to be encountered in typical building surveying activities, e.g. wet and dry rot, flat roof defects, concrete defects, etc.;

- the various methods to collect, store and retrieve information for various differing purposes when carrying out property inspections;

- the various different types of inspection that may be carried out and the importance of accurate recording of information during inspection; and

- differing types of testing and the limitations of the tests, e.g. the use of damp meters and borescopes.

Level 2:

- explain in detail cause and mechanics of varying types of failure;

- explain procedures for carrying out inspections of properties;

- be able to explain, using detailed examples, the relationship between observations taken on site and the diagnosis of failure in building fabric;

- be able to use examples, from your own experience, to demonstrate your application of knowledge gained at level 1; and

- be able to use knowledge and information gathered from several sources, including if necessary specialist inspections to diagnose and explain building fabric failure.

Level 3:

- prepare reports for clients, explaining in non-technical language the causes of failure and the likely results of failure, together with the appropriate remedial measures;

- using information gathered from inspections formulate the necessary remedial/preventative works including specific detail in the form of a schedule of works if required;

- show an understanding of the level of detail required in typical reports, including examples of layout and the use of sketches, drawings and photographs;

- be able to discuss in detail examples of unusual defects you have been involved with and remedial works employed; and

- be able to demonstrate the differing requirements of reports to clients, e.g. the differences between schedules of condition, schedules of dilapidations and pre-acquisition reports.

A special note on the research pathway

Supervisors and counsellors for candidates following the research pathway will note that there are no optional competency requirements. In the special notes for this pathway, it states that candidates are required to demonstrate competence in the research that is applied to the delivery of solutions to a wide range of projects, employing a range of approaches and relating to a number of locations. The research must be relevant to the candidate's chosen pathway. Therefore, for example, if a research pathway candidate is working in a commercial property firm, he or she will need to be signed off to level 1 in five core and optional competencies from the commercial property practice pathway.

There is an additional requirement that the candidate must also demonstrate competence in one competency to level 2 and two competencies to level 1 from the full list of technical competencies.

Conclusion

It is clear from the above that competencies must be selected that enable the candidate to attain the full range of primary skills, guided by the relevant pathway guidance.

Overall, the combination of competencies must reflect the day-to-day work the candidate carries out with his or her employer. The appropriateness of the competency choices will be taken into account at the final assessment interview. The assessors will expect a realistic and sensible choice to have been made, reflecting the skills needed to practise as a surveyor within their chosen field.

WHAT DO THE COMPETENCIES ENTAIL?

The APC guidance sets out the requirements for all competencies at all levels. Each competency is also given a number to allow the candidate to allocate their diarised experience to the relevant competency.

The competencies are set out in a generic way, so that they can be applied to different areas of practice and geographical location. It is important that they are interpreted within the context of the candidate's own area of practice and specialism.

It is obviously impossible to give specific 'one size fits all' guidance on the type of training to be given for each competency. The competencies have been written with a view to flexibility and to allowing the candidate's training to reflect the markets in which they are working.

In general, though, you need to consider providing training and experience which would reflect your firm's expectations of performance by a young graduate after two years at work.

ASSESSING THE COMPETENCIES

The competency achievements confirmed by supervisors and counsellors are recorded in the candidate achievement record.

It is vital that before confirming that a candidate has met any particular level, you carefully study the wording of the particular competency and the further explanation set out in the relevant pathway guidance. You must be satisfied that the candidate has acquired the appropriate experience and will be able to answer questions on the competency at the final assessment interview. Even more importantly, you must be satisfied that his or her achievement in a particular area was not a 'one-off' – you must be confident that they would be able to replicate it at another time, under different circumstances. The point at which a candidate is competent is when you are confident that they could do this, without supervision, to a standard that is acceptable to you. Bear in mind that not only will a candidate not thank you for finding that he or she has been wrongly signed off at too high a level of competence, but you risk lowering the standards of the profession overall.

In deciding at which point a candidate has achieved a particular level of attainment in any of the competencies, there is no minimum number of days of experience required and the number of days of experience needed to reach level 1, 2 or 3 will differ between candidates. The number of days taken to reach the appropriate level will be dependent on a combination of the following factors:

- the starting point – has the candidate any previous experience in the area?;

- the candidate's aptitude and speed of learning in the competency; and

- the quality of the training and experience provided.

Base your judgments on notes you have kept and observations you have made (either directly or indirectly, if your candidate is in another department or on a secondment). You may also rely on evidence and samples of work produced by the candidate. You should, of course, discuss aspects of the competency or level with the candidate, questioning his or her depth of understanding.

With regard to confirming achievement of the competencies, you should think of these as being progressive over the whole course of the training period, with experience being gained over the 23 months for graduate route 1 or 11 months for a graduate route 2 candidate.

The mandatory competencies need to be carefully planned over the 23-month training period. Candidates have ten areas to cover and a balanced approach will be necessary. The conduct rules, ethics and professional practice competency, required to level 3, may cover the full 23 months, whereas more specialist or specific competencies, many of which are only required to level 1, may be achieved in a much shorter period of time.

Always check that the candidate feels confident in the competency at the level being discussed before you confirm achievement. After 23 months, and in the very early part of their career, candidates cannot be expected to know everything. Some of the levels in some of the competencies represent huge areas of knowledge and experience, which may take a long time to acquire.

It is important to recognise that learning in some of these competencies can span a lifetime. You therefore need to be realistic in your expectations as to what can be covered in a 23-month period.

The pathway guidance gives you very useful indications of the experience that is generally expected of a candidate at each level of each competency.

It should also be noted that, in the above examples, although candidates may not have had hands-on experience in a particular area, they are still expected to have some knowledge and understanding of the issues, in order to meet the requirements of the competency.

As a final word of advice, think of your role as that of someone who has the ability to 'stand back' from the day-to-day training process, in order to consider what the candidate has achieved. From his or her point of view, a new candidate may simply have carried out a day's valuation. The supervisor and counsellor must be able to break down that day's work into its constituent factors, and apply these to the competency requirements, to help the candidate understand how that day contributes towards the requirements of the core, optional and mandatory competencies.

Levels 1, 2 and 3

The differences between levels 1 and 2 of each competency are usually fairly straightforward. Level 1 requires in the main, a knowledge and understanding of a particular subject or competency. Level 2 goes on to demand some practical experience – words such as 'to undertake' or 'to carry out' occur in the competency descriptions at this level.

The difference between levels 2 and 3 is a little more complex. Level 2 can be considered to be an application of knowledge or understanding in normal, everyday circumstances. Level 3 aims to extend this application to giving reasoned advice and more in depth technical knowledge. This difference is highlighted in the valuation competency at level 3; the first part of the competency statement calls for candidates to be able to 'demonstrate practical competence in a range of property types or for a range of purposes and to demonstrate the application of a wide range of valuation methods and techniques'. This does not mean that someone with 24 months of experience needs to be able to value an oil refinery or a steelworks – but it does mean that they should have had some experience in a valuation that is more complex than usual. They should also have experience of providing advice to clients.

The competency achievement process should be progressive across the levels. Competency levels should be reached gradually over the training period, reflecting the candidate's gradual progression. However, for graduate route 2 candidates or those who already have surveying experience prior to starting the APC it would be expected that some of the competency levels had been achieved on starting the APC. In these circumstances it may well be the case that a number of competency levels are confirmed by the supervisor and counsellor on the candidate starting their APC. This should be reflected in the structured training agreement for that candidate (see page 55 for more advice on structured training agreements). In addition, the number of days spent reaching a particular level should be reasonable – a candidate is, for example, unlikely to reach level 3 in any particular competency as a result of just six diary training days (unless they have a lot of prior experience in this competency area of course).

It is also important that supervisors adopt a balanced approach to the task, and do not confirm achievement of all of the competencies on the same date. Assessors naturally become suspicious when this occurs.

FINAL ASSESSMENT INTERVIEW: ASSESSING THE COMPETENCIES

In the final assessment interview, not only will the candidate be asked direct questions about particular competencies, but the wider questions asked will enable the assessors to take several competencies into account. The assessors will likewise analyse the documentation received from the candidate, particularly the critical analysis, and will listen to the presentation closely to ensure that the requirements of the competencies have been met.

The competencies will be tested at the level that has been confirmed by the supervisor and the counsellor (which must, of course, at least reflect the minimum requirements for the relevant pathway). So, for example, for a level 1 competency, a candidate might be asked some general questions about inspection, measurement, surveys, or similar. A typical example from the rural pathway would be for a candidate to discuss a valuation of a farm in which he or she was involved. The candidate might be asked to explain how he or she went about the inspection of the farm, its buildings, land and crops.

For level 2, the assessor might question the candidate in some detail as to how he or she prepared the valuation, discussing some of the problems or issues that were encountered. The underlying question for level 2 is: 'How did you apply your knowledge in this area in a practical way?'

Level 3 seeks to extend this further, asking the candidate to look at a particular aspect in a wider context, with questions such as: 'What were the implications of your approach? What other things might have impacted on your actions or this situation?' The assessor will thus step outside of the candidate's direct experience, perhaps posing a theoretical problem. Continuing the rural example, the candidate might be asked how he or she would deal with a valuation if it was discovered that a farm was trading at a loss.

Level 3 aims to extend the questioning beyond the candidate's direct experience, into the wider areas surrounding this experience.

To prepare your candidate for this type of questioning, you should, in your three- and six-monthly reviews, ask similar questions, to encourage a widening of knowledge and understanding. Test your candidate's understanding of the whole work environment, asking questions and setting problems that draw on his or her previous experience, and then stretch this a little further. It is a particularly good idea to employ this approach for some of the mandatory competencies. A candidate with two years' experience, for example, may well have had no personal involvement with an issue relating to ethics or the Rules of Conduct. However, he or she must still have a firm knowledge of what is required in this regard, and the ability to think more widely – and possibly hypothetically – in this area.

More information on the type of questions likely to be encountered by candidates (including examples) is provided in the RICS publication, *APC: your practical guide to success.*

A last word of advice

Competencies are practical, not theoretical. They allow the assessors to ensure that the candidate is competent to practise. Make sure that a candidate's achievement was not a 'one-off'. The candidate must be competent to repeat the achievement under other circumstances.

7 The mandatory competencies

The minimum standards for mandatory competencies are set out in the APC guidance. These competencies are structured in levels.

Your candidate must achieve the minimum standards, as follows:

• Conduct rules, ethics and professional practice	To Level 3
• Client care • Communication and negotiation • Health and safety	To Level 2
• Accounting principles and procedures • Business planning • Conflict avoidance, management and dispute resolution procedures • Data management • Sustainability • Teamworking	To Level 1

In addition, candidates following the senior professional route must achieve the following three competencies:

• leadership (to level 2);

• managing people (to level 2);

• managing resources (to level 2).

The minimum standards described above may also be included at a higher level, if appropriate to the particular pathway. If a pathway includes a mandatory competency to a higher level, this will also appear in the core or optional competency list for that pathway, to the higher level.

When considering your candidates structured training agreement, you should sort out how you propose to help your candidate achieve the required core and optional competencies, and then consider the list of mandatory competencies.

Identify the mandatory competencies where you will provide the appropriate training within your candidates day-to-day work, as part of their structured training. This will leave a remainder which will require you and your candidate to consider additional learning or a specific training course. These actions will qualify for your candidate's professional development.

The mandatory competencies form some of the most important parts of the whole APC. The skills and abilities they encourage and test underpin all professional and technical aspects of working as a surveyor, and are vital for further advancement in the profession.

This chapter will do two things: first, give some further explanation of the requirements for each competency; and second, provide some practical advice on how to help your candidate achieve the required levels, working through the levels outlined. The advice is not intended to be all-encompassing or definitive in any way – each candidate, each pathway, and each organisation will alter the ingredients of each competency slightly. It is certainly not possible to say, 'follow this, and you will pass'. What follows are simply suggestions and pointers as to how the requirements of the competencies might be met. In particular, you should note that the recommended reading is not exhaustive and there are, of course, many other useful sources of reading material that will support these competencies. Each pathway will have its own specialist authors and texts, and you should use your experience to recommend some reading that in your view will be more tailored to your candidate's pathway.

Some of the training suggested – particularly the structured reading – may also prove useful for the candidate's professional development purposes. Remember that some 32 hours of professional development a year must be focused around the areas of the mandatory competencies (16 hours on mandatory competencies and a further 16 hours on professional practice). In particular, you may wish to use the RICS online knowledge resource for property professionals – *isurv* (www.isurv.com). This service, mixing expert commentary with official RICS guidance, covers a huge range of surveying matters, and can be used for professional development purposes.

The minimum requirements for the competencies in the mandatory competency list vary from level 1 to level 3. Remember it is important for your candidate to achieve the required levels in a progressive manner. They cannot just achieve level 3 from a standing start so it is important that they understand the requirements at each level.

CONDUCT RULES, ETHICS AND PROFESSIONAL PRACTICE

The conduct rules, ethics and professional practice competency must be achieved at level 3.

There are two sets of rules: one for members setting out individual obligations and one for firms setting out the firms' obligations, with supporting policies and non-mandatory help sheets as well as additional guidance available for members of the public. APC candidates will need to familiarise themselves with all three aspects to level 3. Copies of the rules and the help sheets can be obtained from www.rics.org/newregulation

Practical guidance

Level 1

The knowledge and understanding required of the candidate at level 1 is wide-ranging. It covers the role, function and significance of RICS; an understanding of society's expectations of professional practice; RICS Rules of Conduct; and the general principles of law and the legal system, as applicable in their area of practice. The Rules are principles-based and candidates should make sure that they are aware of the standards (or principles) that underpin the rules. Other key areas often used in questioning by APC final assessment panels include:

- professional indemnity insurance;

- handling clients' money;

- lifelong learning; and

- avoiding conflicts of interest.

To understand the role and function of RICS, your candidate should carry out some reading around the structure of RICS, the various professional groups, professional group boards and committees, and the functions performed, such as advising the government on housing, taxation, planning and landlord and tenant issues, and bringing an influence to bear on all relevant aspects of society. Apart from keeping abreast of developments in newspapers, a good source of information is the RICS magazine, *RICS Business*, and the various other weekly property publications.

The other requirement at level 1 involves the candidate understanding the role of the professional person and of

society's expectations of such a person. An important
issue is that of ethics, which have been defined as a set of
moral principles extending beyond a formal code of
conduct.

RICS expects members to act both within the Rules of
Conduct and ethically, when delivering surveying services
to clients. As mentioned previously, this approach has
been enunciated as RICS 'principles' and is set out as 12
standards:

1   Act honourably;   Updated   Jan 2011.

2   Act with integrity;

3   Be open and transparent in your dealings;

4   Be accountable for all your actions;

5   Know and act within your limitations;

6   Be objective at all times;

7   Always treat others with respect;

8   Set a good example;

9   Have the courage to make a stand;

10 Comply with relevant laws and regulations;

11 Avoid conflicts of interest; and

12 Respect confidentiality.

Level 2

Level 2 in this competency requires your candidate to
provide evidence of practical application in their area of
practice, being able to justify actions at all times and
demonstrate personal commitment to the Rules of
Conduct, ethics and the associated help sheets.

Level 3

At level 3, your candidate should be able to provide
evidence of the application of the above in their area of
practice in the context of advising clients.

Further guidance

This particular mandatory competency is a very wide topic – but it should be kept in perspective. The final assessment panel will keep their testing and questioning within the confines of the knowledge and experience that a young person, perhaps in their early 20s and occupying a fairly junior position, will have gained. Therefore your candidate will not be expected to have first-hand experience of some of the more detailed areas of the Rules of Conduct for firms, such as your firm's professional indemnity insurance or the details of your clients' money, albeit they will be expected to know what the requirements are around these (and other) areas.

The assessment panel's approach in part will be pitched at some of the areas in which your candidate should have practical experience, such as conflicts of interest and terms of engagement. However, it will also test their wider knowledge around some of the basics, such as why a professional institution has Rules of Conduct, an understanding of the basic principles of professional indemnity insurance and clients' money, and other aspects of the help sheet on maintaining professional and ethical standards. Their training should incorporate a mixture of practical experience, structured reading and perhaps some CPD-type events on current issues. This competency is obviously also an ideal subject for their professional development.

The APC guidance recommends that training in this area accounts for 16 of the 48 hours per year of professional development experience required.

Levels 2 and 3 will involve you developing your candidate's practical experience at work, with an emphasis on taking instructions, understanding and dealing with conflicts of interest, and applying the 12 standards.

At your three-monthly and six-monthly reviews, you should discuss this with your candidate. You may wish to consider three main areas:

- the role and function of RICS: 'What have they read about RICS activities in recent months?'

- the twelve professional and ethical standards: Test your candidate with questions like 'Give me an example of a time at which you have had to know and act within your limitations?'; 'Tell me about an occasion when you had to have the courage to make a stand?'

- Rules of Conduct: 'What have you read this quarter … and what do you understand about the Rules of Conduct concerning clients' money/or similar?'

Some examples of questions asked by assessment panels for this competency are:

- If you are successful with your APC and you decide to start your own business as a chartered surveyor what sort of things will you need to consider and do? For this question, assessors are looking for the candidate to demonstrate that they:
  - will meet the requirements of the code;
  - understand the need for insurances (employers liability, public liability, etc.), health and safety policy, equal opportunities policy, etc.; and
  - will use the RICS logo as set out by RICS and use only the appropriate alternative designation relevant to their professional group.

- What is client's money and how can a firm preserve the security of this?

- If you are approched in respect of an instruction and agree terms over the telephone what would you then need to do and what details would you need to provide?

- If you are successful in undertaking a project for a client and they send you a case of wine as a thank you, what should you do?

- You are working for a surveying practice and a friend asks you to … [example of work from the relevant professional group] … as a favour because you are a surveyor, what should you do?

You will see from the above examples how the questions tend to be scenario based and will often take a candidate into areas in which they may not have direct experience.

The RICS Regulation website (www.rics.org/newregulation) provides an excellent source of information regarding the RICS Rules of Conduct, including details of disciplinary hearings and minutes of the meetings of the Regulatory Board.

Finally, you should ensure that your candidate can provide evidence of this competency in their experience record and professional development record – giving examples of structured reading and training and also of practical experience. Candidates must also ensure that all their detailed activities clearly comply with the Rules of Conduct and associated standards.

The DVD *RICS Rules of Conduct* available from RICS Books is an excellent resource for candidates for this competency. The DVD is designed to inform and challenge the candidate's knowledge and raise awareness of the depth and the breadth of the Rules of Conduct and what is expected of a professional.

CLIENT CARE

This competency is required to be achieved to level 2. At level 1, your candidate should be able to demonstrate a knowledge and understanding of the principle and practice of client care, including:

- the concept of identifying clients, colleagues and third parties who are your clients and understand the behaviours that are appropriate to establish good client relationships;

- the systems and procedures that are appropriate for managing the process of client care including complaints; and

- the requirements to collect data, analyse and define the needs of clients.

At level 2, they should be able to apply all of the above in their area of business or practice on a routine basis.

Practical guidance

Approaches to client care will vary from business to business, depending on the nature of the work, the degree of client interface and the type of organisation. Candidates must be able to understand the link between customer care and duty of care. If a client wishes to do something that would be impractical, or impossible, or doomed to certain failure, the surveyor owes the client a duty of care to inform him or her of that. To give the best customer care, your candidate must therefore have a good understanding of each client's needs.

This is one of those competencies where the candidate must be able to 'step back' out of a situation, to analyse what it is that they have learnt about customer care and duty of care in any particular instance. In preparation for the final assessment, they should be able to explain a situation in which they have delivered good client care in the context of their work. You should discuss this with your candidate and provide opportunities where they can develop skills in this area.

To assist their understanding of this subject, the candidate may also wish to undertake some structured reading and training.

It is also essential that your candidate is familiar with RICS complaint handling procedures and how to confirm instructions, including the items to include within terms of engagement.

COMMUNICATION AND NEGOTIATION

This competency must be achieved at level 2.

At level 1, candidates are required to demonstrate a knowledge and understanding of effective oral, written, graphic and presentation skills, including methods and techniques appropriate to specific situations. At level 2, they are required to provide evidence of practical application of oral, written, graphic and presentation skills that are appropriate in a variety of situations, specifically where negotiation is involved.

Candidates will be able to demonstrate oral communication skills from a wide range of surveying situations and circumstances: at meetings, in negotiations, when managing people, when making presentations, in tenders, and so on. It should include the use of email and of internal memos and letters, for all of which an essential component is being able to write good, unambiguous prose.

Communication skills can be taught formally. A wealth of bodies run courses in this area – these may include, for example, assertiveness training courses. However, the best approach may be for candidates to be coached by senior people in their particular area, and to put their developing skills continually into practice.

On a general level, any course or training programme should cover the nature and purposes of oral communication – addressing the different approaches to be taken in different situations, and the techniques that can be used to communicate effectively.

A search on the RICS Books website (www.ricsbooks.com) for 'communication skills' will identify a number of excellent publications that candidates can use to hone their skills for this competency.

Remember that the final assessment interview includes a ten-minute presentation. This is therefore one of the easiest of the mandatory competencies for the panel to assess.

Practical guidance – communication skills

For level 1, your candidate must be able to understand the various media in which written communications can be presented, and more importantly, the skills involved in doing so, with regard to the target audience, the length, style and layout of the communication, the message they wish to convey, and the structure of the communication. In terms of 'graphic' communication, it covers sketch notes, drawings in plans, designs linked to the construction process and similar (if these are relevant to the candidate's chosen pathway). It is probably easiest to assess understanding with reference to actual written work – their own and other people's – in a variety of mediums. They should consider why a particular communication fails or succeeds, how it could be improved, and what the aspects are that make it successful in a particular area.

For level 2, candidates need to show that they can put this knowledge into practice. They will need to consider a range of their written work, in a variety of media. Is it all appropriate for the audience and purpose? Have they achieved, in a letter, memo, report, email, sketch or design what they set out to achieve? If they are experiencing difficulties in this area they could attend a course (internal or external) on written communication. Practice will be key – nobody would expect them to write a perfect client report the first time they tried, but with a full understanding of the principles and purposes of the report, and some more practice, they will be much better equipped to do so.

The critical analysis submitted by your candidate for the final assessment provides an easy opportunity for the panel to assess the candidates' level of competence in this area.

Practical guidance – negotiation skills

This competency overlaps somewhat with the 'conflict avoidance' competency, and the section on that competency should prove useful here too (see page 100).

To fulfil the requirements of this particular competency, check that your candidate understands what lies behind successful negotiations: the preparation of evidence; an understanding of the various approaches to negotiations; a knowledge of where and how parameters are set; a knowledge of what each side wishes to get from the negotiations, and from any future relationship; and so on. If possible, involve your candidate in negotiations. This is perhaps the best way of helping them to gain an understanding of principles and skills.

HEALTH AND SAFETY

This competency is required to level 2.

The basic level 1 requirement is to demonstrate a knowledge and understanding of the principles and responsibilities with regard to health and safety imposed by law and codes of practice and other regulations relating to health and safety appropriate to your candidate's area of practice. At level 2 candidate's will be required to provide evidence of practical application of health and safety issues and the requirements for compliance, in their area of practice.

Practical guidance

This competency covers all aspects of a surveyor's working life. It is about ensuring that the surveyor's entire working life is conducted as safely as possible with as little risk to health as possible, and that the same is true for all of those around the surveyor. It is easy to think of ways in which health and safety issues relate to, say, work on a construction site, but perhaps less so for more office-based work. However, the same basic philosophies underpin all work carried out in any environment. In off-site jobs, the issues encompass such things as managers knowing where a candidate is and what they are doing at all times, and, should they leave the office, when they will return and who they are meeting. There are also numerous health and safety issues relating to the use of equipment, in offices as well as all other locations, and on keeping employees healthy and safe.

Owing to the importance of health and safety, most firms and organisations conduct formal training and instruction on the relevant issues. Ensure that your candidate attends this and that they can explain the reasons behind any requirements imposed by the firm.

Structured reading will no doubt be needed to cover the range of issues for this competency. The Health and Safety Executive (HSE) provides numerous free leaflets on its website at www.hse.gov.uk, including lists of its current publications. You should encourage your candidate to visit the site and select some useful reading.

Finally, they must be able to demonstrate knowledge of the health and safety legislation and codes of practice that apply to their area of work, and also evidence of practical application as required by level 2 of the competency. This will differ from one pathway to another, but includes such things as the *Health and Safety at Work Act* 1974, the *Construction (Design and Management) Regulations* 2007 and the *Control of Asbestos Regulations* 2006. Candidates must be aware of relevant legislation and be able to explain it and its significance to their area of work.

ACCOUNTING PRINCIPLES AND PROCEDURES

This competency must be achieved to level 1. The requirement is to demonstrate knowledge and understanding of fundamental accounting concepts and the format and preparation of management and company accounts, including profit and loss statements, cash flow statements and balance sheets.

Practical guidance

Your candidate will undoubtedly be able to gain some experience – practical or theoretical – of these concepts in the course of their work. If there is one particular aspect that is unlikely to arise in their everyday work, consider how training might be arranged so that this is covered. If this proves impossible you should encourage your candidate to read a textbook on the subject, or attend an appropriate training course or CPD lecture.

It should be noted that some candidates might quite easily reach levels 2 and 3 in this competency if this subject is part of their job. For example, a rural practice candidate working for an estate would deal with these types of issues regularly. Candidates working in commercial property and dealing, perhaps, with the leisure and entertainments industry, might also easily attain level 2 or 3 through practical experience and use of accounts in undertaking valuations and in assessing the strength of potential tenants.

BUSINESS PLANNING

This competency must be achieved to level 1. The requirement is to demonstrate knowledge and understanding of how business management activities contribute to the achievement of corporate objectives.

Practical guidance

This is a vast subject, and candidates will generally benefit by attending some kind of basic management training course, where possible. These are run by a number of bodies, including Open University and the Chartered Management Institute. The candidate may well not be directly involved in managing your business, and this is not expected of them for the purposes of level 1, but they must be able to understand – and explain to the assessors, if necessary – the underlying factors of business management.

A lot will also be learned on the job, of course, and coaching and training from you will be invaluable. There is a lot of overlap with the requirements of the other mandatory competencies (communications and negotiation, teamworking, client care, and so on). There are a range of books available that will help to develop knowledge and understanding of issues such as motivation, mission statements, strategy, organisational structures, and so on.

A useful text book in this respect is *The handbook of management and leadership a guide to managing for results* by Michael Armstrong and Tina Stephens (Kogan Page, London, 2005). This book covers the practice of management, delivering change, enhancing customer relations and enabling continuous improvement.

CONFLICT AVOIDANCE, MANAGEMENT AND DISPUTE RESOLUTION PROCEDURES

This competency must be achieved to level 1. This requires knowledge and understanding of the techniques for conflict avoidance, conflict management and dispute resolution procedures, including for example adjudication and arbitration appropriate to your candidate's APC pathway.

Practical guidance

The 'ingredients' of this competency will vary greatly between the various APC pathways. In commercial practice, for example, landlord and tenant matters of conflict will be fairly common, while in construction, this competency will be present everyday in managing building contracts. Indeed, in the quantity surveying and construction pathway, this becomes a core competency to level 2, with requirements based around procurement and the drafting of terms and conditions of leases, contracts and agreements.

In basic terms, and across all pathways, it is important that the candidate understands how to conduct negotiations, and also the various options available should negotiations break down, working through mediation and conciliation, adjudication, arbitration, independent expert determination, and, finally, litigation.

Candidates should be encouraged to sit in on negotiations at their firm from an early stage in their career. Also they will benefit from some formal training on this and other aspects of dispute resolution, covering the preparation of evidence, case law, approaches and tactics. It is reasonably likely that by the time they reach the final assessment, candidates will need to have had practical experience of running their own negotiations, or participating in other dispute resolution procedures, and will thus be able to discuss this. As part of their training plan ensure that they make steady progress towards this end.

There are also many texts available on this subject. The following RICS guidance notes and practice statements will be useful background reading:

- *Surveyors acting as arbitrators and as independent experts in commercial property rent reviews*;

- *Surveyors acting as adjudicators in the construction industry*;

- *Surveyors acting as expert witnesses*; and

- *Surveyors acting as advocates*.

All of these publications are available in hardcopy from RICS Books (www.ricsbooks.com). RICS members can download the publications as a PDF from www.rics.org

In addition, don't forget CPD-type lectures or training that may be available within the firm, or from external providers.

DATA MANAGEMENT

This competency must be achieved to level 1. It involves demonstrating knowledge and understanding of the sources of information and data applicable to the candidate's area of practice, including the methodologies and techniques most appropriate to collect, collate and store data.

Practical guidance

Again, this competency will vary greatly between the APC pathways. It is important to think of it in relation to the candidate's specific route, and against the backdrop of their day-to-day work and the particular IT developments in their area.

In the valuation pathway, for example, this competency will cover comparable evidence found in sales and rental evidence. Collection, collation and storage methods in this pathway will usually comprise the use of IT spreadsheets and databases, either developed by firms or sold as commercial packages. Developments in this area, and in the commercial property pathway, include the use of computer assisted techniques.

In the quantity surveying and construction pathway, meanwhile, sources of data may be previous contracts or cost guides and price books. Various commercial packages are also available to price contracts and bills of quantities. For all pathways, the important thing is for the candidate to be able to understand the use of data in their day-to-day work – how this is gathered and put to use, and what the best methods of collection, collation and storage are. Candidates should be able to step back mentally from their work, to explain what data they use, how they find it and how it is manipulated. It is also important for candidates to be aware of the implications of data protection legislation and how this will affect the use that can be made of data that organisations may hold.

You should avoid the temptation to 'write this competency off', on the basis that it will be covered at level 1 elsewhere – for example, in level 1 of the valuation competency. Try to use the competency to broaden and develop the candidate's understanding of wider data issues and developments in the profession. See this competency as a subject in itself and ensure your candidate carries out some structured reading. Discuss it as a discrete issue at some point in their training plan and at the three- and six-monthly review stages.

SUSTAINABILITY

As a mandatory competency this is required to level 1. It requires a knowledge and understanding of why and how sustainability seeks to balance economic, environmental and social objectives at local, national and global levels, in the context of land, property and the built environment.

Practical guidance

All chartered surveyors need a basic understanding of environmental issues, which range from groundwater pollution and contaminated land, to control of pollution in the air we breathe, and on to even wider global issues, such as climate change. Environmental issues affect building design, construction use and management, development and re-development, and regeneration and town planning. Issues such as global warming, dwindling national resources and atmospheric pollutants are top priorities with many government and influential bodies, such as the EU and the World Trade Organisation (WTO).

Candidates should carry out some general reading in newspapers and professional journals on environmental issues. Other useful sources of information and advice are RICS responses to government consultations and the RICS guidance notes and information papers, including:

- *Carbon management of real estate*;

- *Contamination and environmental issues – their implications for property professionals*;

- *Renewable energy*; and

- *Sustainability and the RICS property lifecycle.*

Surveying Sustainability: a short guide for the property professional, produced in partnership between RICS,

103

Forum for the Future and Gleeds and the sustainability pages on the RICS website (visit www.rics.org/ sustainability) are also useful.

In addition, of course, candidates should maintain an awareness of environmental issues whilst at work. Candidates should be aware of any government initiatives, laws or EU regulations affecting their particular area of work. On a more local level, they should be aware of any internal office environmental policies (recycling of paper, for example), and be able to explain the purposes of these. You may like to test their knowledge and understanding in this area by asking them to consider how they would express the firm's 'green credentials', should this be requested in, for example, an invitation to tender. Once more, this is a case of stepping back from day-to-day work, for them to consider the environmental factors that underlie and overarch such work.

TEAMWORKING

This competency is required to level 1. Candidates must demonstrate a knowledge and understanding of the principles, behaviour and dynamics of teamworking.

Practical guidance

This competency involves understanding why people work in teams, and some of the basic principles underlying teamworking. In practice, candidates will rarely not work within a team, so in effect, while the mandatory requirement is to level 1, in practice they will often be working to level 2 ('working as a team member in a work or business environment').

Evidence of working in a team will be easy to come by; however, the level 1 requirement is to understand the principles behind this. Candidates should therefore consider a situation in which they have witnessed or experienced teamworking and be able to explain how that team worked, concentrating on the roles each member adopted and the success or otherwise of this.

Candidates understanding can be complemented and extended by some reading on the subject. A leading text on this subject is *Management teams: Why they succeed or fail*, by R. M. Belbin, (Butterworth-Heinemann, Oxford, 2003).

A last word of advice

Hopefully you now have a better idea of what the mandatory competencies entail, and of how to help your candidate achieve them. Remember that the philosophies behind the mandatory competencies, and the business skills inherent in them, will be encountered in every aspect of their working life. It is for this reason that they are mandatory.

8 Referred candidates and appeals

We come now to the very last stage in the APC process. Your candidate has attended the final assessment interview and is waiting for the decision.

The assessment panel will make a decision as to whether your candidate has passed or will be referred. Normally, the outcome will then be notified to the candidate within 21 days. If the result is a pass, then all is well and good, and you can skip the rest of this chapter. If, however, the outcome is a referral (fail), the notification will include a referral report, which will give guidance as to why the panel has reached this decision.

Referral will be a miserable experience for the candidate, and it is important that the supervisor and the counsellor are on hand to provide help. In the first instance you will need to provide support by acting (in the usual sense of the word) as a counsellor, and may therefore wish to arrange a meeting with the candidate. In preparation for this, ask the candidate to write a few notes to reflect his or her memory of the experience, with reference to the contents of the referral report.

At the meeting, the first thing you will need to do is let the candidate 'get it off their chest'. You should sit back and listen, adopt an understanding tone and offer some words of comfort and a shoulder to cry on. When the more emotional stage is over, ask the candidate to discuss how he or she feels about the outcome, and centre your discussion on the referral report, checking the candidate's understanding of the reality of the situation. At all times remain realistic – at the end of the meeting you will be looking to agree one of two outcomes: to make an appeal, or to live with the referral and apply to take the final assessment again.

You should strive to achieve an outcome in which your candidate accepts responsibility for the situation, regardless of whether you intend making an appeal. This should provide a positive way forward for all concerned and put your candidate in the right frame of mind for the future – a win-win situation all round.

If you feel the referral is justified, then it is time to start planning how to succeed next time. If, however, the candidate wishes to make an appeal, then he or she has ten working days (from the date the result is posted by RICS) to do so.

Details of how to make an appeal are available on www.rics.org. Broadly speaking, appeals may be made on one of three grounds:

- administrative or procedural: the panel may not have been provided with the correct information or detail;

- the questioning and testing of competence concentrated too heavily outside of the candidate's main areas of training and experience; or

- any form of discrimination.

In most instances no appeal will be lodged. The next step for the supervisor and the counsellor is therefore to steer the candidate through (at least) a further six months of training and to make preparations for the next assessment. It is important that a detailed action plan is put in place to help focus this on the necessary areas. A review of the structured training agreement may be necessary. RICS training advisers can provide help and support for supervisors and counsellors who are putting plans in place for referred candidates.

There are some minimum requirements that referred candidates must satisfy. They must:

- undertake additional training and work experience;

- ensure that the deficient competencies outlined in the referral report have been signed off again by the supervisor and counsellor when the deficiencies identified have been fully addressed;

- undertake a minimum of a further 24 hours of professional development;

- write a new critical analysis or if recommended by the assessors, resubmit the original, suitably amended, with updates;

- agree with the supervisor and counsellor how the deficiencies identified in the referral report will be addressed; and

- submit the referral deficiency report and professional development deficiency report.

The advice given previously in this book on filling in the experience record (see chapter 5) should be followed for the referral deficiency report, which is a very similar document and simply requires the candidate to demonstrate the further experience that they have gained.

At the end of the six months, or later, if the candidate has deferred further, the candidate will be re-interviewed. This will take the same format as the original interview, including a presentation on the relevant critical analysis.

It is of course important that the supervisor and counsellor continue to provide support and encouragement over the six-month period. You will be involved in planning the additional training and in assisting the candidate to fill in the referral deficiency record. You will need to carry out another six-monthly-style review, checking all relevant documentation and assessing the competencies that were considered deficient by the APC assessment panel.

WHAT WENT WRONG – AND HOW DO WE PUT IT RIGHT?

At this stage, the candidate, supervisor and counsellor will want to know 'what went wrong' the first time round? Knowing this, they can endeavour to put it right for the next time. Most referred candidates pass at the second attempt – making it clear that there is a simple way of adjusting or adding to performance, to ensure success.

Having said that, there is no magic formula to achieve a pass at the second attempt. However, the following examples indicate where a concentration of effort in one particular area may be the key to success. The referral report will of course indicate which specific aspects should be considered.

Sometimes, there is something very concrete and easy to put right. Often, this relates to the critical analysis. You should make sure that the new report submitted matches all the requirements set out in chapter 5 regarding the critical analysis and in the APC guidance.

Another omission that, in a final assessment situation, will be an instant referral, is a lack of knowledge of the Rules of Conduct. Make sure your candidate has a good knowledge of these for their second attempt – see page 87 for full advice on these.

Sometimes, the first failure was just about underpreparation in general, often in combination with overconfidence. Go back over the stages in this book, test and retest the candidate on the competencies, and have them re-present their presentation to you and another colleague. Make sure they know everything back to front. Check also that they have a good, wider knowledge of each area, and have not simply learnt something verbatim, leaving them unable to explain the principles or processes behind an action. This can be the case where candidates use software to undertake activities, for example valuations or costings, and are not aware of the basic principles behind this. Candidates must be able to do the same job with a piece of paper!

On other occasions, a candidate may simply have given a poor performance on the day or may have suffered with nerves. Ask them to undergo a mock interview with you, or another supervisor or colleague. Assess their performance under pressure and consider how it could be improved.

Sometimes, the candidate has just had back luck – with particular problems hard to define. Often, the solution in these instances is simply to undertake a little more training and experience.

In the experience of senior assessors, the reasons for referral can be broken down roughly as follows. Some 85% of candidates fail simply because they do not demonstrate that they are not yet sufficiently competent. As noted above, to achieve competence, it is often just a matter of improving experience and knowledge in one particular area; a lack of competence at this stage is by no means an indication that it will never be achieved. The referral report will indicate any particular areas of weakness to focus on.

Another 10% of candidates are referred because they fail to demonstrate competence. Their critical analysis may have been weak, their presentation poor, or the examples they produced to support their achievements poorly thought through. If you know your candidate is competent, but if he or she failed to get this across at the interview, then consider, using the referral report, how he or she can better demonstrate competence in the problematic areas.

The final 5% fail for a variety of reasons: by making simple mistakes on the day, often caused by a bad case of nerves; by making mistakes in the submitted documents (which checking and double-checking can avoid); or as a result of other, usually avoidable problems, which are unlikely to be repeated a second time round.

With the pass rate for second attempts at the APC being very high, candidates should certainly not despair; neither should their supervisors or counsellors. A little more time and effort is almost always all that is required.

A final word of advice

Having helped one particular referred candidate to subsequent success, supervisors and counsellors should take some time to consider whether the lessons learnt could be applied to other candidates in the system.

9 Help and advice from RICS

We hope that this book has given you an insight into the importance of your role as a supervisor or counsellor – and sufficient advice on how to carry it out. The roles do carry with them a great deal of responsibility; however, you are not alone.

RICS has set up extensive help facilities for candidates and employers. The following points of assistance can help with any of the aspects of your role outlined in this book:

● RICS Contact Centre;

● RICS website;

● Regional representatives – the RICS regional teams and Matrics often organise training sessions for APC candidates. Contact details for regional administrators can be obtained from RICS (see page 125);

● RICS training advisers (RTAs); and

● APC doctors.

Remember too, that advice is provided in the official APC guidance and the pathways guidance. The RICS publication, *APC: your practical guide to success*, will also prove useful.

RICS TRAINING ADVISERS (RTAs)

RICS employs training advisers for each region in the UK. Details may be found on the RICS website at www.rics.org/apc

The role of the RTAs is to advise organisations on all aspects of the APC. Most importantly, they are able to help firms develop structured training agreements and to provide approvals for these (see chapter 5 for more information on these). In drawing up such an agreement, the RTA will provide initial advice to the firm. When the firm has prepared a training agreement that meets with the minimum standards described in chapter 5, the RTA will confirm approval and it may then be used as an indicator of excellence and of the firm's commitment to APC training – which may prove very useful in job advertisements and at recruitment fairs.

Over the years RTAs have built up a wealth of knowledge and understanding of APC training across the APC pathways. As such, they are an excellent source of guidance and knowledge.

APC DOCTORS

Whereas the primary role of the RTAs is to support and advise employers, voluntary APC doctors are available to assist and guide candidates through the system. They are normally locally based and, where practical, will be from the same professional group as the candidate's chosen route. They are often recently successful APC candidates and can therefore provide others with the benefit of their first-hand experience. Details of local APC doctors can be found at www.rics.org/apc

READING MATERIAL

All of the guidance notes and practice statements noted in this book can be obtained from www.ricsbooks.com

RICS Books can likewise assist with the ordering of the other publications listed in this book. APC candidates may find particularly useful the RICS online knowledge resource for property professionals – *isurv* (www.isurv.com). The isurv 'channels' cover a huge range of surveying topics, mixing expert commentary with official RICS guidance. The *isurv* APC channel also holds excellent support material (www.isurv.com/apc)

DISCUSSION FORUM

The APC discussion forum is an excellent opportunity for candidates to ask questions and share experiences with one another. This can be accessed from the members' area on the RICS website.

Conclusion

The candidates that arrive at your firm are your raw material. You, the supervisors and counsellors, must shape them into future surveyors, which can seem a daunting task. Hopefully, the advice in this book, and the additional guidance and support available from RICS, will ensure that you feel comfortable in carrying out that task successfully.

The following thoughts may help you with this task:

Respect the individuality and diversity of your candidates. Do not expect candidates to be the mirror image of yourself – or imagine that this would necessarily be a good thing!

Remember also that different candidates require different approaches. If a project has not gone too well, a straightforward approach, with (constructive) criticism, may work for some people. For others, you may have to employ a greater level of tact. If you take the time to get to know your candidates, you will soon understand which approach works best for them.

Keep your candidate, and his or her progress, constantly in your mind. Do not think of the APC process as something to dip in and out of at specified intervals. Rather, you should be always on the lookout for opportunities for your candidate, chances to resolve difficulties, and prospects for the future.

In one large firm, a supervisor became aware of major negotiations taking place in another department. Aware that her candidate could gain much from these, she persuaded her colleagues to allow the candidate to sit in on these, at a much earlier stage in his career than would otherwise have been the case.

Far and away the most common call for help to RICS comes from candidates who feel that their supervisors and counsellors are not spending enough time with them, or taking enough of an interest in their progress. There is obviously a simple answer to such problems – spending more time, or, if this is not possible, communicating the reasons for this to the candidate.

Understand the importance of 'standing back'. To be able to identify your candidates' achievements against the competencies, you need to have the ability to stand back from their day-to-day work, and consider how the requirements of the competencies are reflected in that. This will help you explain how to fill in the various records of progress, thinking of each day's work in terms of its constituent factors. You should also encourage your candidate to acquire the same skill of 'standing back', so that they can answer unexpected questions at the final assessment interview. Candidates who understand the principles and processes behind everything they do will be able to answer any question that is thrown at them.

Avoid the 'halo and horns' effect, which refers to what happens when objectivity gets lost. If you have had a good experience with a candidate, you may be ready to believe the best of them in all situations, and may avoid pointing out failings, preferring not to mention these. After a bad experience, on the other hand, it can be easy to view a candidate over-critically. The two experiences often go hand in hand, with one candidate appearing to do no wrong, and another appearing to do no good. Strive hard to resist either extreme. Make sure you look objectively at your candidates' work, and seek confirmation of your judgments from other colleagues.

Act professionally at all times. You are of course steering your candidates through a process in which they must learn to abide by the RICS code of ethics and Rules of Conduct. Throughout the training process you will be a walking, living example of these Rules in practice. You must therefore act professionally towards your candidates at all times, leading by example.

Do not put undue pressure on your candidate to take the final assessment. People do progress at different speeds. The worst thing that can occur is to enter an underprepared candidate for the final assessment or a candidate who is significantly lacking in confidence of their own ability. The subsequent referral gets the candidate's career off to a bad start, and makes everyone feel dissatisfied. Two years is a minimum target – there is no shame in taking longer.

If your candidate is not ready for the assessment (even if he or she thinks they are), or alternatively, if you feel they are ready, but they think otherwise, then listen to them, and consider how you can reach an agreed state of readiness. You may wish to arrange a meeting, at which strengths and any 'underdeveloped' strengths can be identified, and a clear timetable drawn up, indicating how skills will be gained. Clarity should provide confidence.

Seek advice if problems arise which you feel you cannot resolve. This is particularly pertinent if problems arise in which you yourself are involved. We are only human – it is not unknown for relationships to break down on occasion. In these circumstances, act carefully and thoughtfully. Have you contributed, in any sense, to the breakdown or difficulties? How could you modify your approach or behaviour? Seek advice from your peers who may have encountered similar situations or alternatively discuss the matter with senior colleagues (where relevant you could also try the RICS Membership Operations department).

Above all, remember that as a supervisor or counsellor you have someone looking up to you, requiring your guidance and expertise. By following the advice in this book you will be very well-equipped to guide your candidate to success!

Index

six-monthly reviews 17,
18
three-monthly reviews
16, 27, 28
change of employer form
55
client care 92–94
communication skills
94–96
competencies 72–73
a competency 72
assessing 79–83
final assessment
interview 83–84
choosing 75–78
entailment 78–79
list and levels 73–74,
81–83
mandatory *see*
mandatory
competencies
types 74–75
competency-based
assessment 23–24,
72, 73
competency-based
interviews 10
conduct rules, ethics and
professional practice
87–92
level 1 88–89
level 2 89
level 3 89
conflict avoidance,
management and
dispute resolution
procedures 100–101

*Construction (Design and
Management)
Regulations* 2007 98
continuing professional
development (CPD)
62
continuous assessment
22–23
*Control of Asbestos at
Work Regulations*
2006 98
core competencies 74, 75
counsellors *see* supervisors
and counsellors
counsellor's progress report
18, 36–37
critical analysis 68–71
actions at 12 months
38–39
actions at 18 months 40
actions at 21 months 40
actions at 23 months 43
mark sheet 49–50
one month after
submission of final
assessment
application 19, 20
referrals 109

data management 101–102
deferrals 41, 108
definitions
a competency 72
diary 62–63
actions at 23 months 42
supervisor actions
six-monthly reviews 17,
18

RICS contact details and further information

There are numerous ways of contacing RICS and a wealth of information sources online for APC candidates.

RICS Contact Centre

T: +44 (0)870 333 1600

F: +44 (0)20 7334 3811

E: contactrics@rics.org

RICS website

The RICS website provides details on the routes to membership, how to enrol and how to apply for the final assessment. You can also download the APC guidance, templates, Excel workbooks and pathway guidance and find your nearest regional training adviser and APC doctor. Visit www.rics.org

RICS Books

T: +44 (0)870 333 1600

F: +44 (0)20 7334 3851

E: mailorder@rics.org

www.ricsbooks.com

isurv

isurv is an online knowledge resource for property professionals, brought to you by RICS. The *isurv* 'channels' cover a huge range of surveying topics, mixing expert commentary with official RICS guidance. Visit www.isurv.com

The *isurv* APC channel also holds excellent support material and provides links to information relevant to achieving each competency. Visit www.isurv.com/apc